“十三五”高等教育机电类专业规划教材

电气自动化技术专业英语

Electrical Automation Technology English

姚　薇　钱玲玲　主编

中国铁道出版社有限公司
CHINA RAILWAY PUBLISHING HOUSE CO., LTD.

内 容 简 介

本书以提升电气自动化技术、机电一体化技术等相关岗位专业英语技能为目标，以学生容易看懂的英文图纸、软件界面、设备操作面板、设备说明书、企业官网与产品宣传手册为教学内容，使专业英语学习简单化、实用化，有助于学生进行本专业技能的拓展与提升。

本书主要内容包括基本电路、集成电路、电气控制技术应用、传感检测技术应用、PLC、变频器、气动系统设备、电力系统装置、智能控制、机床、典型机电控制设备（机器人、电梯）等。

本书采用由浅入深的方式安排内容，循序渐进地引导学生进行专业英语的学习，以掌握必要知识和学习方法，从而促进学生学习能力的提升。

本书适合作为高等职业院校电气自动化技术、机电一体化技术等专业的教材，也可作为相关专业人员的参考书。

图书在版编目（CIP）数据

电气自动化技术专业英语 / 姚薇，钱玲玲主编 .—2 版 .—北京：中国铁道出版社有限公司，2020.8 （2024.1重印）
“十三五”高等教育机电类专业规划教材
ISBN 978-7-113-27166-4

Ⅰ．①电… Ⅱ．①姚… ②钱… Ⅲ．①电气化 - 自动化技术 - 英语 - 高等学校 - 教材 Ⅳ．① TM92

中国版本图书馆 CIP 数据核字 (2020) 第 147413 号

书　　名：电气自动化技术专业英语
作　　者：姚　薇　钱玲玲

策　　划：王春霞　　**编辑部电话：**（010）63551006
责任编辑：王春霞　绳　超
封面设计：付　巍
封面制作：刘　颖
责任校对：张玉华
责任印制：樊启鹏

出版发行：中国铁道出版社有限公司（100054，北京市西城区右安门西街 8 号）
网　　址：http://www.tdpress.com/51eds/
印　　刷：河北宝昌佳彩印刷有限公司
版　　次：2013 年 8 月第 1 版　2020 年 8 月第 2 版　2024 年 1 月第 3 次印刷
开　　本：787 mm×1 092 mm 1/16　**印张：**11　**字数：**263 千
书　　号：ISBN 978-7-113-27166-4
定　　价：32.00 元

FOREWORD 前言

本书以提升电气自动化技术、机电一体化技术等相关岗位专业英语技能为目标确定教学模块，结合技术发展动态和技术综合应用，组织英文内容，具有较强的专业针对性；以学生能够看懂的英文图纸、软件界面、设备操作面板、设备说明书、企业官网与产品宣传手册为教学内容，使专业英语的学习简单化、实用化。本书最大限度地吸纳项目化的教学理念,实现了专业技术和专业英语的有机融合。全书从简单的电工电子元件、仪表设备入手，涉及工业控制设备的产品手册、系统构造规则等，以图片的形式呈现，简单易学。

本书分为 6 章，共 19 个单元。电路理论基础模块主要是以图文并茂的形式介绍元器件、仪表、参数等英文表述；电气控制系统、工业控制系统模块兼顾专业基础理论知识，培养学生理解设备的安装和阅读产品手册的能力；电力系统保护模块主要使学生掌握电力保护装置的产品名称；机电一体化技术模块主要使学生了解典型机电设备各部件的英文描述；拓展部分选取了相关阅读材料，可提高学生阅读英文文献的能力。

本书相较于第一版在内容选取上涉及自动应用的多个领域，增加了机器人、电梯这些典型的自动化设备的内容，同时介绍了其行业的优秀企业。

教师可根据专业实际和课时数自由选取合适的单元组织教学内容。

本书由姚薇、钱玲玲主编，并负责全书内容的选取与构架安排。王超、薛岚在本书的编写中提出了宝贵意见，特此感谢。

由于编者水平有限，书中难免存在不足和疏漏之处，敬请读者批评指正。

编　者

2020 年 2 月

CONTENTS 目录

CHAPTER 1 Circuit Theory Fundamentals

CHAPTER 2 Electric Control System

CHAPTER 3 Industrial Control System

CHAPTER 4 Power System Protection

Unit 10 Power System Protection

CHAPTER 5 Mechatronics Technology

Unit 11 Mechatronics

Unit 12 Machine Tools

Unit 13 Elevator

Unit 14 Robot

CHAPTER 1

Circuit Theory Fundamentals

Objectives:

(1) Explain how the current flows.
(2) List some basic electronic components.
(3) Know each component in a power supplies.
(4) List the application of electronics.
(5) Differentiate between analog and digital electronics.
(6) Describe different logic gates.

Unit 1

Electrical Circuitry

1. Three basic properties used in electrical circuits

All components used in electrical circuits have three basic properties, known as capacitance, inductance and resistance, as shown in Figure 1.1. In most cases, however, one of these properties will be far more prevalent than the other two.

1) Resistance

That characteristic of a medium opposes the flow of electrical current through itself.

2) Capacitance

The property of a component is to oppose any change in voltage across its terminals, by storing and releasing energy in an internal electric field.

3) Inductance

The property of a component is to oppose any change in current through itself, by storing and releasing energy in a magnetic field surrounding itself.

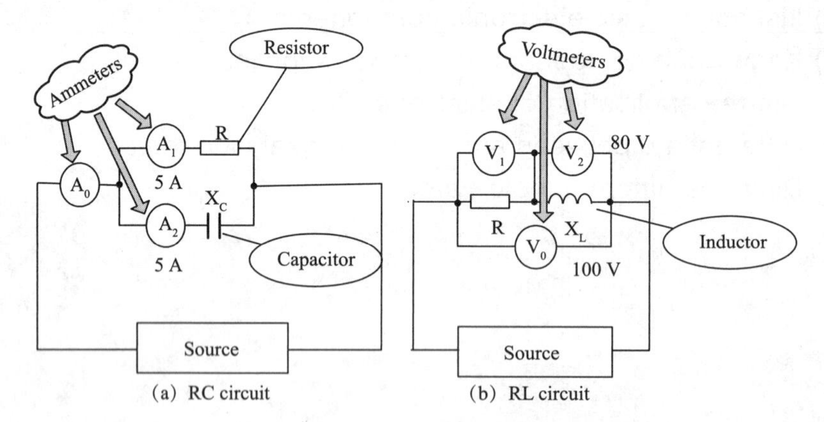

Figure 1.1 Basic circuits

As you might expect, components whose main property is resistance are called resistors; those that exhibit capacitance are called capacitors, and the ones that primarily have inductance are called inductors.

1) Resistor

As you could probably guess from the name, a resistor increases the resistance of circuit. The main purpose of this is to reduce the flow of electricity in a circuit. Resistors come in all different shapes and sizes, as shown in Figure 1.2.

2) Capacitor

A capacitor is a component made from two (or two sets of) conductive plates with an insulator between them. The insulator prevents the plates from touching. The strength of a capacitor is called capacitance and is measured in Farads (F). They are used in all sorts of electronic circuits, especially combined with resistors and inductors, and are commonly found in PCs, as shown in Figure 1.3.

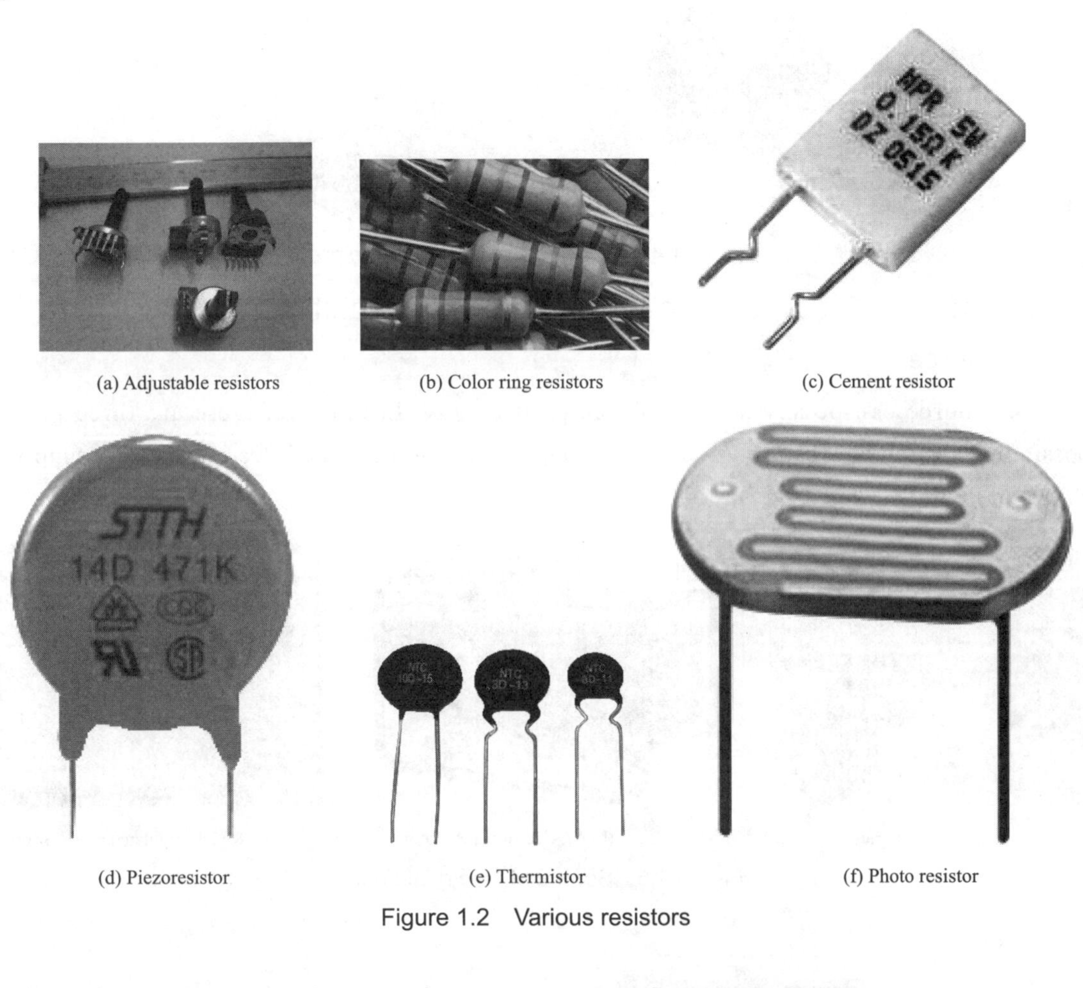

(a) Adjustable resistors (b) Color ring resistors (c) Cement resistor

(d) Piezoresistor (e) Thermistor (f) Photo resistor

Figure 1.2 Various resistors

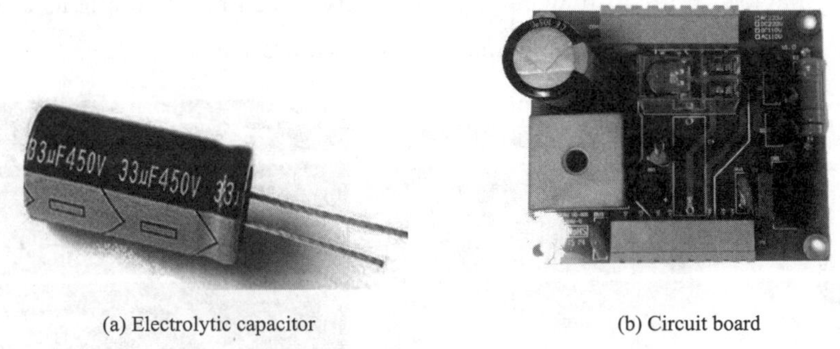

(a) Electrolytic capacitor (b) Circuit board

Figure 1.3 Application of the capacitor

3) Inductor

An inductor is essentially a coil of wire. When current flows through an inductor, a magnetic field is created, and the inductor will store this magnetic energy until it is released. In some ways, an inductor is the opposite of a capacitor. The strength of an inductor is called, take a wild guess, its inductance, and is measured in Henrys (H). A picture of different inductors as shown in Figure 1.4.

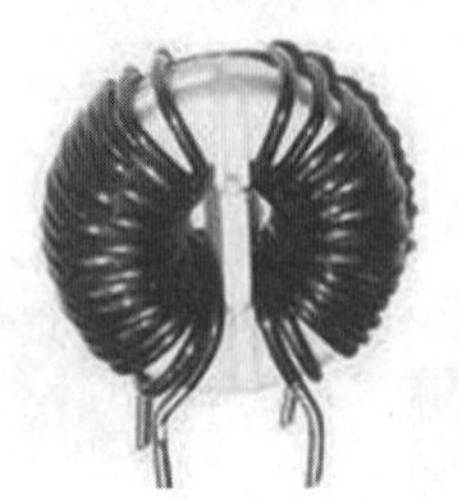
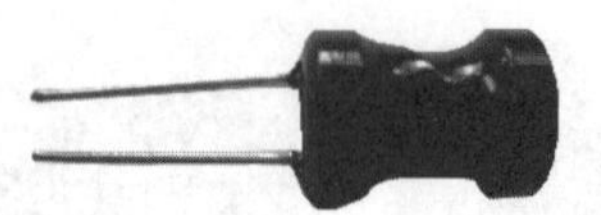

Figure 1.4 A picture of different inductors

2. The consist of a basic type of electric circuit

1) Source

The source can be any source of electrical energy. In practice, there are three general possibilities: it can be a battery, an electrical generator, or some sort of electronic power supply, as shown in Figure 1.5.

(a) Batteries

(b) An electrical generator

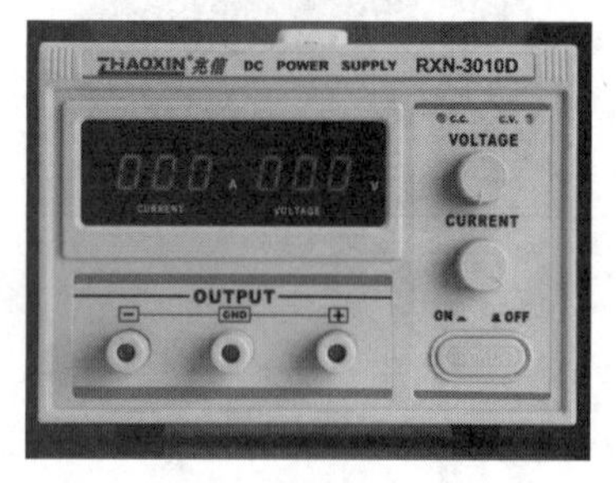

(c) DC regulated power supply

Figure 1.5 Source of electrical energy

2) Load

The load is any device or circuit powered by electricity. It can be as simple as a light bulb or as complex as a modern high-speed computer, as shown in Figure 1.6.

(a) A light bulb

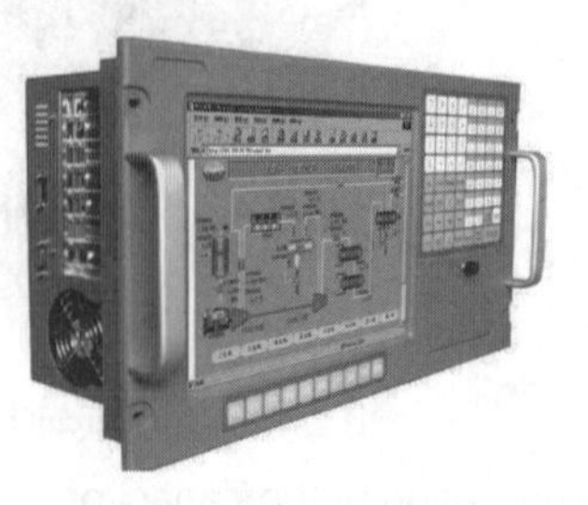

(b) Industrial control equipment

(c) A modern high-speed computer

Figure 1.6 Load for different purposes

3. The physical characteristics of the circuit

The electricity provided by the source has two basic characteristics, called voltage and current. These are defined as follows:

1) Voltage

The electrical "pressure" that causes free electrons to travel through an electrical circuit, is known as electromotive force (EMF). It is measured in volts by voltmeter, as shown in Figure 1.7.

Figure 1.7 Voltmeters

2) Current

The amount of electrical charge (the number of free electrons) moving past a given point in an electrical circuit per unit of time is called current. Current is measured in amperes by ammeter, as shown in Figure 1.8.

Figure 1.8 Ammeter

The load, in turn, has a characteristic called resistance. The characteristic of a medium opposes the flow of electrical current through itself. Resistance is measured in ohms by ohmmeter, as shown in Figure 1.9 and Figure 1.10.

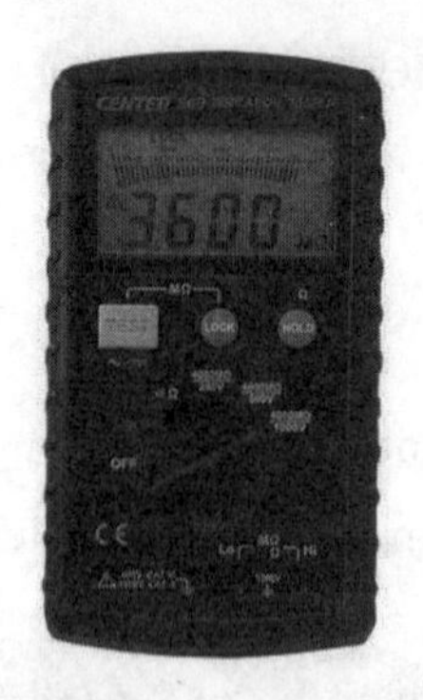

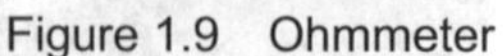

Figure 1.9 Ohmmeter　　Figure 1.10 Multimeter

Extension Materials

Attention in the use of electrolytic capacitor

(1) If unclear or unknown of polarity of circuit indication. Non polar capacitor is recommended to be used.

(2) Electrolytic capacitors should not be used at the environmental temperature which exceeding the range of specified operating temperature.

(3) Electrolytic capacitors should be stored in low temperature and dry place, if stored in long time, re-aging process with rated voltage must be taken prior to use capacitor.

(4) Heaters of high temperature, such as electric irons, should be kept away from the vinyl insulation sleeve of capacitor which to avoid the cause made over heat or break down to sleeve.

(5) If excessive force is applied to the lead wires or terminals, the inside construction of capacitor should be destroyed.

电解电容器使用注意事项

（1）在电路回路中如不清楚或不明确线路的极性时，则建议使用无极性电解电容器。

（2）电解电容器的工作环境温度不能超过规定的使用温度范围。

（3）电解电容器应储存于低温及干燥场所，如储存期较长，则使用前应用额定电压对其重新校验。

（4）电烙铁等高温发热装置应与电解电容器塑料外壳保持适当的距离，以防止过热造成塑料套管破裂。

（5）对导线、端子，如施加超过规定的力，将会破坏电解电容器的内部结构。

Self-Test

(1) Fill in the blanks with the information in the reading material.

① Components whose main property is resistance are called___________; those that exhibit capacitance are called________; and the ones that primarily have inductance are called________.

② The electricity provided by the source has two basic characteristics, called__________and __________.

③ Electromotive force (EMF) is measured in__________; current is measured in__________; resistance is measured in______.

(2) Find the right description and connect to it.

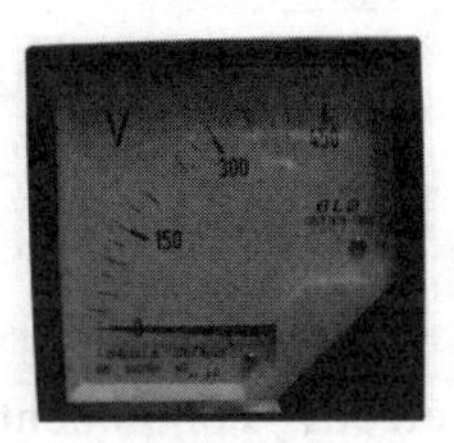

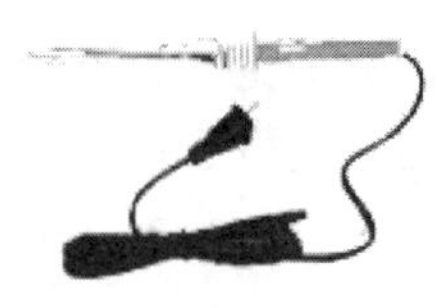

voltmeter

electric iron

ammeter

multimeter

New Words and Phrases

resistance *n.* 阻力；电阻，热阻；抗力；抵抗力

resistor *n.* 电阻器

capacitance *n.* 电容，电容量

capacitor *n.* 电容器；电容

inductance *n.* 电感；感应系数；自感应

inductor *n.* 电感器

piezo-resistor *n.* 压敏电阻器

photoresistor *n.* 光敏电阻器，光电导管；光电阻器

electrical current 电流

voltage *n.* 电压

electrolytic capacitor 电解电容器

load *n.* 负荷；工作量，负荷量

multimeter *n.* 万用表

electron *n.* 电子

source *n.* 源极；信号源；电源

bulb *n.* 球茎；电灯泡

Volt *n.* 伏[特]（电压单位）

Ampere *n.* 安[培]（电流单位）

Ohm *n.* 欧[姆]（电阻单位）

voltmeter *n.* 伏特计；电压表

ammeter *n.* 安培计；电流表

Unit 2

Electronics

1. Something about electronics

The world's reliance on electronics is so great that commentators claim people live in an "electronic age". People are surrounded by electronics-televisions, radios, computers, and DVD players, along with products with major electric components, such as microwave ovens, refrigerators, and other kitchen appliances, as well as hearing aids and medical instruments.

Many different methods of connecting components have been used over the years. For instance, early electronics often used point to point wiring with components attached to, wooden breadboards to construct circuits. Cordwood construction and wire wraps were other methods used. Most modern day electronics now use printed circuit boards (made of FR4), and highly integrated circuits. You can see the four types of construction methods in Figure 2.1.

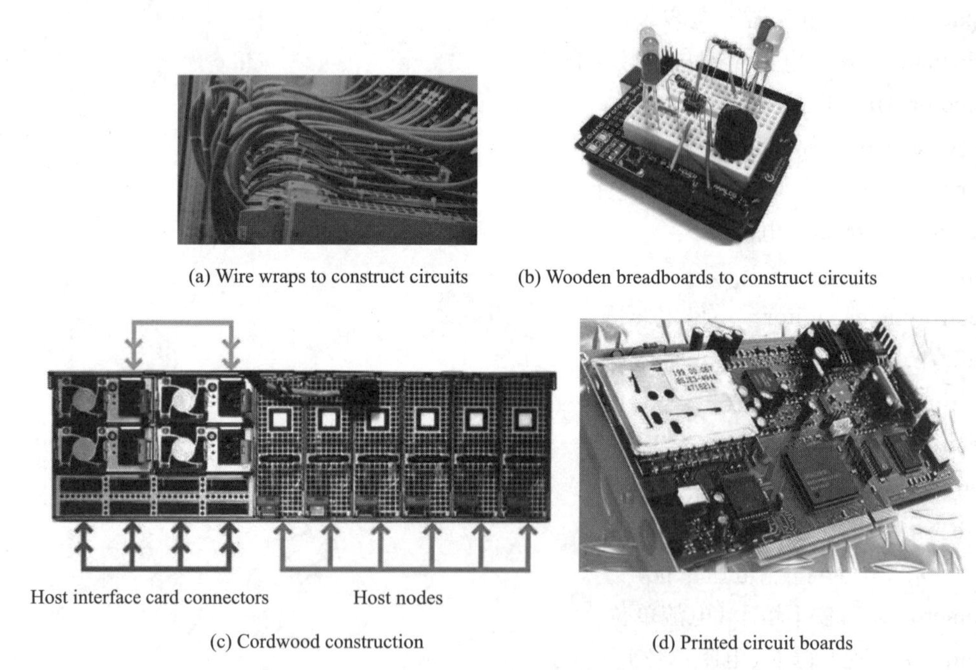

(a) Wire wraps to construct circuits (b) Wooden breadboards to construct circuits

(c) Cordwood construction (d) Printed circuit boards

Figure 2.1 Four types of construction methods

Circuits and components can be divided into two groups: analog and digital. A particular device may consist of circuitry that has one or the other or a mix of the two types.

2. Analogue electronics

Analogue electronics are electronic systems with a continuously variable signal.

The term “analogue” describes the proportional relationship between a signal and a voltage or current that represents the signal.

1) Electronic components with analog switch ICs

Analog switch ICs, as shown in Figure 2.2(a), have slowly turned on designers for their efficiency and very popular feature. IC analog switch integrated circuit that electric current flows, if allowed, restricted and closed when opened.

The analog (or analog switch), also called the bilateral switch, as shown in Figure 2.2(b), is an electronic component that behaves the same way as a relay, but has no moving parts. Analog switch ICs for commercial, industrial, as used for military purposes because they are widely used in various electronic switches IC analog components. The IC switches are also used for analog signals to digital control interface.

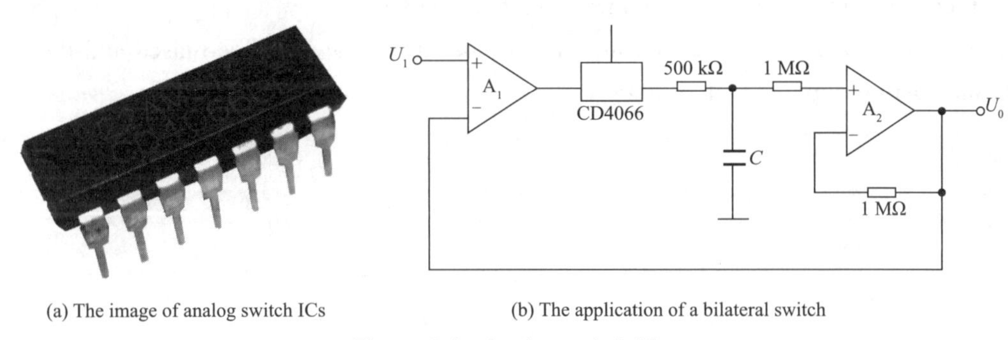

(a) The image of analog switch ICs (b) The application of a bilateral switch

Figure 2.2 Analog switch ICs

Analog switches are used in various applications such as mobile phone, PDA, digital camera, laptop, LCD TV and set-top box.

2) Amplifiers

An amplifier can be considered as an electronic circuit with an input port to which a signal enters, and an output port from which an enlarged signal emerges.

Many kind of amplifying devices are available, and most are adaptable for use in control system. If the output of the amplifier delivers more power than is required at the input, the amplifier is generally classes as a power amplifier. Another class of amplifier is the voltage amplifier.

Amplifiers in electrical systems may be as simple as an electromagnetic relay or they may consist of electronic devices such as tubes, transistors, or magnetic circuits.

Amplifiers and electronic devices is shown as Figure 2.3.

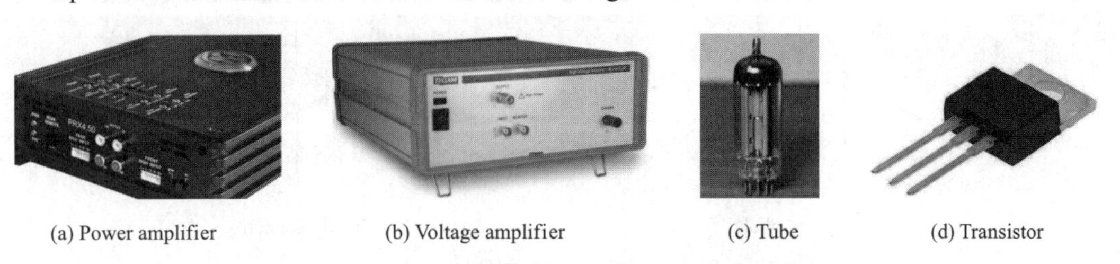

(a) Power amplifier (b) Voltage amplifier (c) Tube (d) Transistor

Figure 2.3 Amplifiers and electronic devices

3. Digital electronics

A digital circuit is often constructed from small electronic circuits called logic gates (see Figure 2.4-Figure 2.6). Each logic gate represents a function of Boolean logic. A logic gate is an arrangement of electrically controlled switches. The output of a logic gate is an electrical flow or voltage that can, in turn, control more logic gates.

The basic AND gates can be made very simply using diode and resistors. Figure 2.4(b) shows a two-input AND gate made with diodes and resistor, which works in the following manner.

If both inputs are connected to a low voltage (logic 0), current flows from the positive supply through the resistor and the two diodes in parallel back to ground. The output is held to the voltage drop across the conducting diodes, or at a low (logic 0) level.

If one of the inputs is connected to a logic 1 high level, that diode is reverse-biased, but the other still conducts to ground, keeping the output at a low level.

If both inputs are connected to a high level, both diodes are reverse-biased and the output rises to the positive supply level, or logic 1.

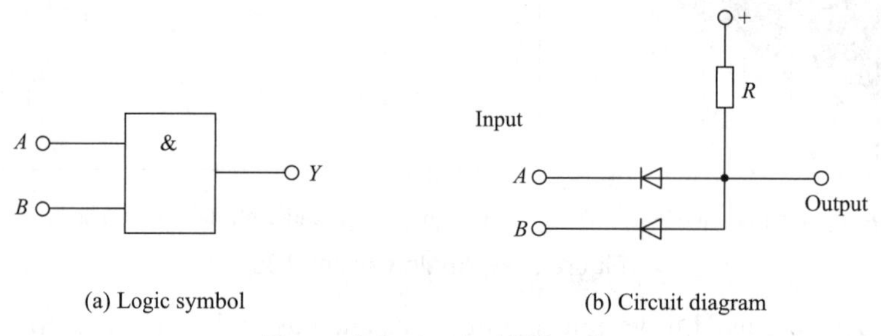

Figure 2.4 Logic AND gate

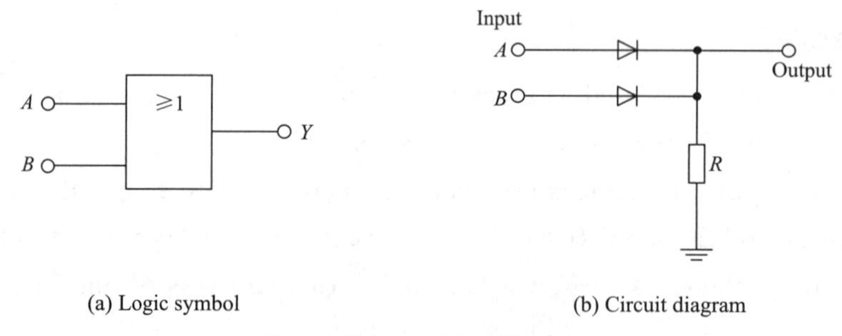

Figure 2.5 Logic OR gate

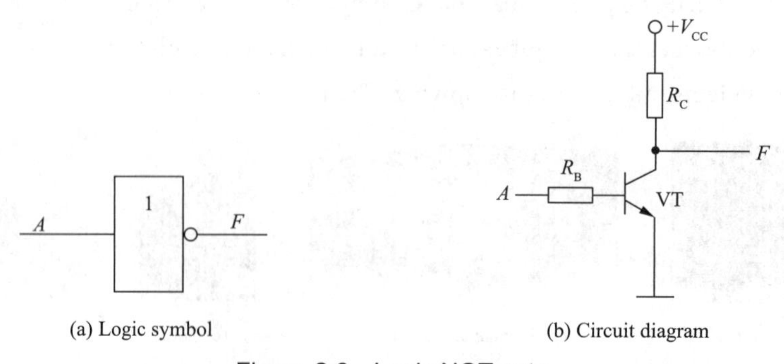

Figure 2.6 Logic NOT gate

Extension Materials

Carbon microphone

One of the first devices used to amplify signals was the carbon microphone (effectively a sound-controlled variable resistor). By channeling a large electric current through the compressed carbon granules in the microphone, a small sound signal could produce a much larger electric signal. The carbon microphone was extremely important in early telecommunications; analog telephones in fact work without the use of any other amplifier. Before the invention of electronic amplifiers, mechanically coupled carbon microphones were also used as amplifiers in telephone repeaters for long distance service.

Magnetic amplifiers

A magnetic amplifier is a transformer-like device that makes use of the saturation of magnetic materials to produce amplification. It is a non-electronic electrical amplifier with no moving parts. The bandwidth of magnetic amplifiers extends to the hundreds of kilohertz.

碳素麦克风

碳素麦克风是第一批用于放大信号的设备之一(实际上是一种声音控制的可调电阻器)。通过麦克风中的压缩碳颗粒引导一个大电流，一个小的声音信号可以产生一个更大的电信号。碳素麦克风在早期通信中极其重要，实际上，模拟电话不用任何其他放大器就能工作。在发明电子放大器之前，机械耦合碳素麦克风也被用于电话中继器的长途业务中。

磁 放 大 器

磁放大器是利用磁性材料的饱和产生放大的类似变压器的装置。它是一种无运动部件的非电子电放大器。磁放大器的带宽可延伸到几百千赫。

Self-Test

As shown in Figure 2.5(b), that shows a two-input OR gate made with diodes and resistor. Fill in the blanks with the state information of the circuit according to the input signal.

(1) If both inputs are connected to a low voltage (logic 0) ______________________________

____

The output ____

(2) If one of the inputs is connected to a high level (logic 1) ____________________________

____

The output____

(3) If both inputs are connected to a high level__________________________________

____

The output ____

New Words and Phrases

electronics *n.* 电子学；电子学的应用；电子器件
digital *adj.* 数字式的；数码的；数字信息系统的；数位的
integrated *adj.* 整体的；完全的；综合的；整合的
proportional *adj.* 比例的，成比例的
medium *n.* 媒介，手段，方法，工具
ICA 集成通信适配器
IC 集成电路
bilateral *adj.* 双边的，双方的
resistance *n.* 电阻
laptop *n.* 便携式计算机
LCD 液晶显示器
set-top box 机顶盒

CHAPTER 2

Electric Control System

Objectives:

(1) Describe the English name for various types of motors.
(2) Recognize the nameplate of motor.
(3) Know the AC motor control system.
(4) Know the DC motor control system.

Unit ③

Electric Motors

Each type of motor, as shown in Figure 3.1, has its particular field of usefulness. Because of its simplicity, economy and durability, the induction motor widely used for industrial purposes than any other type of AC motor, especially if a high-speed drive is desired.

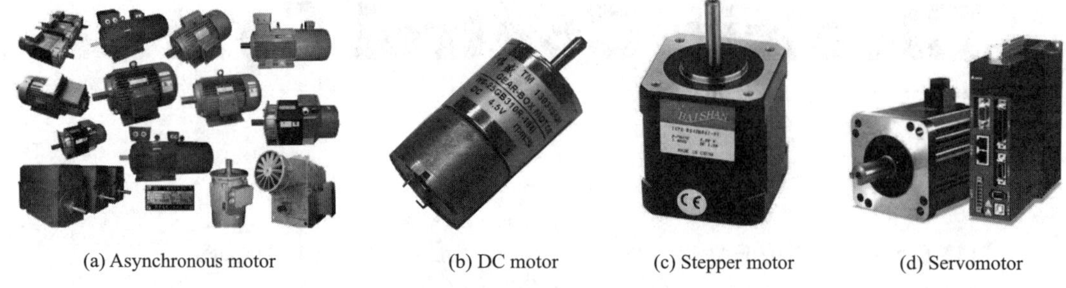

(a) Asynchronous motor (b) DC motor (c) Stepper motor (d) Servomotor

Figure 3.1 Various types of motors

If AC power is available, all drives requiring constant speed should use squirrel-cage induction or synchronous motors because of their ruggedness and lower cost. Drives requiring varying speeds, such as fans, blowers, or pumps, may be driven by wound-rotor induction motors. However, if there are machine tools or other machines requiring adjustable speed or a wide range of speed control, it will probably be desirable to install de motors on such machines and supply them from the AC system by motor-generator sets or electric rectifiers.

Almost all constant-speed machines may be driven by AC squirrel-cage motors because these motors are made with a variety of speed and torque characteristic. When large motors are required or when power supply is limited, the wound-rotor motor is used, even to drive constant-speed machines.

For varying-speed service, wound-rotor motors with resistance control are used for fans, blowers, and other apparatus for controller and duty and are used for cranes, hoists, and other installations for intermittent duty. The controller and resistors must be properly chosen for the specific application. Synchronous motors may be used for almost any constant-speed drive requiring about 100 hp (1 hp=745.700 W) or over.

Cost is an important factor when more than one type of AC motor is applicable. The squirrel-cage motor is the least expensive AC motor of the three types considered and requires very little control equipment. The wound-rotor is more expensive and requires additional secondary control. The synchronous motor is even more expensive and requires a source of DC excitation, as well as special synchronizing control to apply the DC power at the correct instant. When very large machines are involved as, for example, 1,000 hp or over, the cost picture may change considerable and should be checked on an individual basis.

The various types of single-phase AC motors and universal motors used very little in industrial applications, since poly-phase AC or DC power is generally available. When such motors are used, they are usually built into the equipment by the machinery manufacturer, as in portable tools, office machinery, and other equipment. These motors are, as a rule, especially designed for the specific with which they are used.

1. Speed control of DC motor

The internal structure of DC motor is shown in Figure 3.2. The connection of DC motor control system is shown in Figure 3.3.

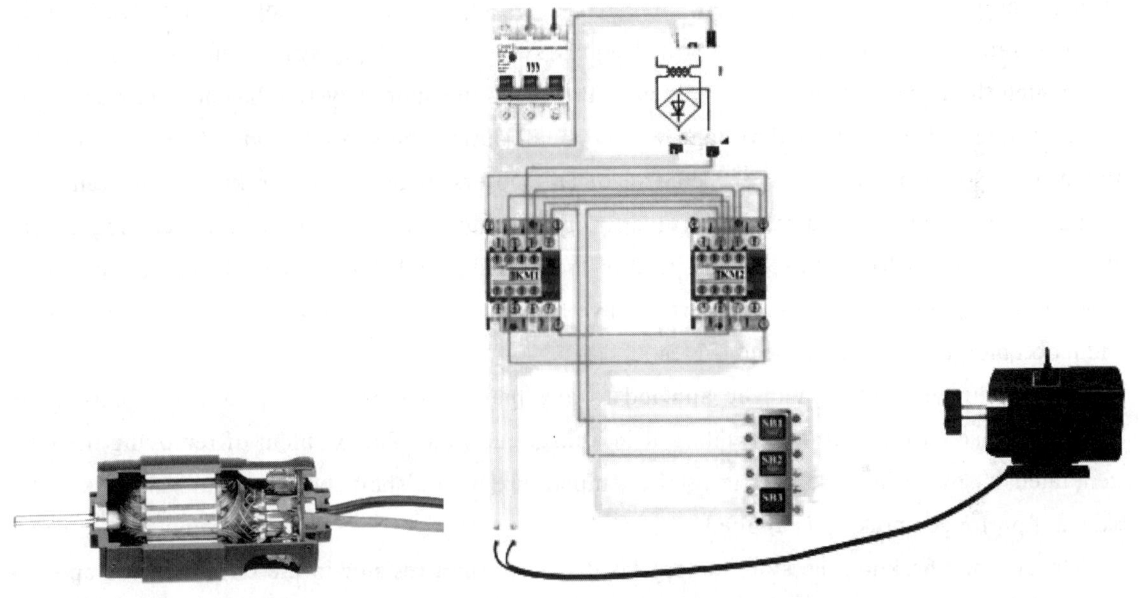

Figure 3.2 The internal structure of DC motor   Figure 3.3 The connection of DC motor control system

1) A regulator system

A regulator system is one which normally provides output power in its steady-state operation. For example, a motor speed regulator maintains the motor speed at a constant value despite variations in load torque. Even if the load torque is removed, the motor must provide sufficient torque to overcome the viscous friction effect of the bearings. Other forms of regulator also provide output power: A temperature regulator must maintain the temperature of, say, an oven constant despite the heat loss in the oven. A voltage regulator must also maintain the output voltage constant despite variation in the load current. For any system to provide an output, e.g., speed, temperature, voltage, etc., an error signal must exist under steady-state conditions.

2) Electrical braking

In many speed control systems, e.g., rolling mills, mine winders, etc., the load has to be frequently brought to a standstill and reversed. The rate at which the speed reduces following a reduced speed demand is dependent on the stored energy and the braking system used. A small speed control system can employ mechanical braking, but this is not feasible with large speed controllers since it is difficult and costly to remove the heat generated.

The various methods of available electrical braking are:

(1) Regenerative braking.

(2) Eddy current braking.

(3) Dynamic braking.

(4) Reverse current braking (plugging).

Regenerative braking is the best method, though not necessarily the most economic. The stored energy in the load is converted into electrical energy by the work motor (acting temporarily as a generator) and is returned to the power supply system. The supply system thus acts as a "sink" into which the unwanted energy is delivered. Providing the supply system has adequate capacity, the consequent rise in terminal voltage will be small during the short periods of regeneration. In the Ward-Leonard method of speed control of DC motors, regenerative braking is inherent, but thyristor drives have to be arranged to invert to regenerate. Induction motor drives can regenerate if the rotor shaft is driven faster than speed of the rotating field. The advent of low-cost variable-frequency supplies from thyristor inverters have brought about considerable changes in the use of induction motors in variable speed drives.

Eddy current braking can be applied to any machine, simply by mounting a copper or aluminum disc on the shaft and rotating it in a magnetic field. The problem of removing the heat generated is severe in large system as the temperature of the shaft, bearings, and motor will be raised if prolonged braking is applied.

In dynamic braking, the stored energy is dissipated in a resistor in the circuit. When applied to small DC machines, the armature supply is disconnected and a resistor is connected across the armature (usually by a relay, contactor, or thyristor). The field voltage is maintained, and braking is applied down to the lowest speed. Induction motors require a somewhat more complex arrangement, the stator windings being disconnected from the AC supply and reconnected to a DC supply. The electrical energy generated is then dissipated in the rotor circuit. Dynamic braking is applied to many large AC hoist systems where the braking duty is both severe and prolonged.

Any electrical motor can be brought to a standstill by suddenly reconnecting the supply to reverse the direction of rotation (reverse current braking). Applied under controlled conditions, this method of braking is satisfactory for all drives. Its major disadvantage is that the electrical energy consumed by the machine when braking is equal to the stored energy in the load. This increases the running cost significantly in large drives.

2. Frequency control for AC motor

Figure 3.4 shows the structure of a three-phase asynchronous motor. It contains stator, rotor, winding, etc.

Figure 3.4 Three-phase asynchronous motor

Let's review for a moment the concepts of line and forced commutation as they are used to obtain adjustable frequency to be applied to an AC squirrel cage induction motor. Figure 3.5 illustrates what the controller is supposed to do: create an adjustable voltage and adjustable frequency from fixed line voltage and fixed line frequency. Let's first decide what we need to put in the box. The circuit using six thyristors will not work. It can create an adjustable voltage to the motor, but the line frequency passes straight on through. Therefore, we need a means of creating an adjustable frequency as well as an adjustable voltage. The simplest way of doing this is by means of a "DC link". The DC link is then controlled by one of several means to create the adjustable frequency. In some cases the DC link is also controlled to create the adjustable voltage. To form the DC link, the incoming AC voltage must somehow be changed to a DC voltage, after which the DC is changed back to AC for applying to the AC motor.

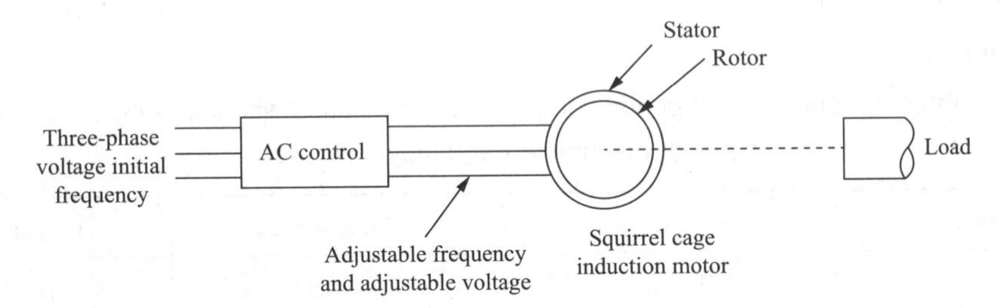

Figure 3.5 Conversion functions of adjustable-frequency controllers applied to AC motors

Figure 3.6 shows a generalized frequency controller with a DC link. The input uses six semiconductors to provide the DC, and the output uses six semiconductors to provide the adjustable frequency. Which type of semiconductor should be selected for each box? Since the AC line is always connected to the first box, the input devices can be line commutated. They can therefore be diodes, thyristors, GTOs, transistors, or triacs. GTOs are quite costly, and transistors and triacs may not have the desired ampere capacity and voltage capabilities. Therefore, the input power devices will be diodes or thyristors or perhaps a combination.

Again the devices in the output box of Figure 3.6 must utilize forced commutation because there is no natural or line means to turn off the power semiconductors. This means that they must be thyristors, transistors, or GTOs. Triacs could be used but are limited in capacity.

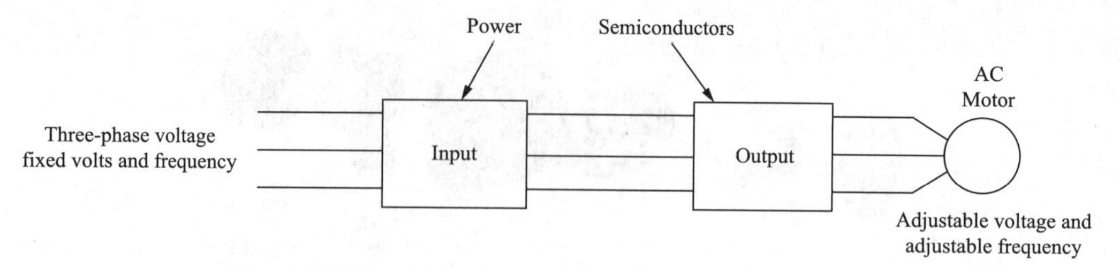

Figure 3.6 Generalized adjustable-frequency controller with DC link

Let's start with diodes in the input box. With diodes, there is no means of adjusting the DC link voltage. Therefore, both the adjustable frequency and the adjustable voltage must be created in the output stage. This is actually done in the real case. The resulting system is called pulse-width modulation (PWM).

If the input box were to use thyristors instead of diodes, they can be controlled to provide adjustable DC link voltage. The output stage then needs only to create the adjustable frequency from the DC link and pass the adjustable voltage on through to the AC motor, together with the adjustable frequency. The output stage can therefore be less complex than a PWM system, but the input stage must be more complex. However, this method is also in popular use for frequency controllers, in one of two arrangements. One is known as the adjustable voltage inverter (AVI) . The other is known as the current source inverter (CSI).

Table 3.1 is a summary of power devices as used in these three most common types of frequency controllers.

Table 3.1 Summary of power semiconductors as used with the three basic types of adjustable-frequency controllers

Type of controller	Input devices	DC link voltage	Output devices
PWM	Diodes	Constant	Thyristors
AVI	Thyristors	Adjustable	Diodes
CSI	Thyristors	Adjustable	GTOs

1) Terminology

Let's review for a moment some of the various terms as they are used to describe solid-state frequency controllers. There are rigid technical definitions as well as generally used terminology. First, the technical definitions as suggested by organizations such as IEC, NEMA, and IEEE (Institute of Electrical and Electronics Engineers):

(1) Converter: an operative unit for electronic power conversion comprising one or more valve devices (power semiconductors, for example).

(2) Self-commutated converter: a converter in which the commutation voltages are supplied by components within the converters.

(3) Rectifier: a converter for conversion from AC to DC.

(4) Inverter: a converter for conversion from DC to AC.

(5) Indirect AC converter: a converter comprising a rectifier and an inverter with a DC link.

If we look at Figure 3.6, all definitions above apply to solid-state frequency controllers in one form or another. However, general usage in the United States is to call the configuration of Figure 3.6 an "inverter": PWM inverter, adjustable voltage inverter, or current source inverter [also called an adjustable-current inverter (ACI)]. For the rest of this lesson we use the terms "adjustable-frequency controller" and "inverter" interchangeably, recognizing, of course, that we can have the whistle blown on us at any time for not using correct technical terminology.

2) PWM versus AVI versus CSI

All three of the most commonly used adjustable-frequency controllers consist of three basic sections, as shown in Figure 3.6. The input section converts the incoming AC power to DC. The center section, or DC link, smoothes out or filters the DC voltage. The output section inverts the DC into AC of the desired frequency.

The differences among these three types of controllers are the manner in which the adjustable voltage is obtained; the technique used to create the adjustable frequency.

3) PWM

Pulse-width modulation (PWM) utilizes diodes in the input stage to provide a fixed-voltage DC bus. The output, or inverter stage, creates a series of pulses of constant voltage with the pulse widths and pulse quantities varying as required by the desired output frequency and voltage. The output section supplies and controls both parameters, adjustable frequency and adjustable voltage.

4) AVI

Adjustable-voltage inverters (AVIs) use thyristors in the input stage to obtain adjustable voltage in the DC link. The output stage switches this DC voltage with thyristors or transistors or GTOs to obtain a square-wave voltage whose width and timing sequence are proportional to the desired frequency. Voltage control is obtained in the first stage. Frequency control is obtained in the second stage.

5) CSI

Current source inverters (CSIs) are similar to AVIs, except that the control is arranged to provide a series of square waves of current output.

6) Comparison of PWM, AVI, and CSI

Figure 3.7 shows the three types of adjustable-frequency controllers, with power circuits and resulting theoretical output voltage and current waveforms. There are numerous variations within these three basic systems, such as sine-wave modulation with PWM, chopper techniques, and output circuitry that provides a form of load commutation. These refinements are beyond the scope of this lesson and are covered in the references.

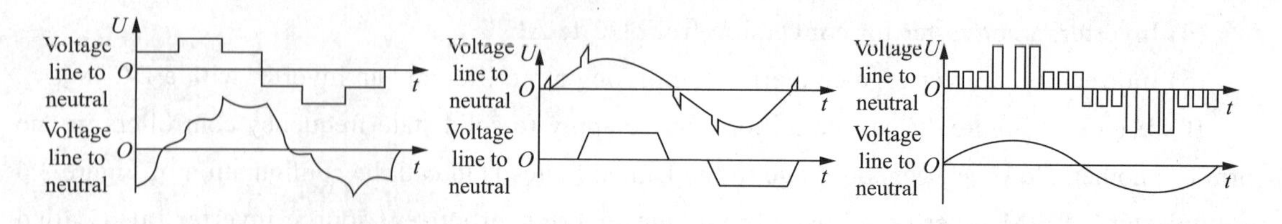

Figure 3.7 Power circuit and output waveforms of basic types of adjustable-frequency controllers

Extension Materials

A single-quadrant speed control system using thyristors

A single-quadrant thyristor converter system is shown in Figure 3.8. For the moment the reader should ignore the rectifier BR2 and its associated circuitry (including resistor R in the AC circuit), since this is needed only as a protective feature and is described in next section.

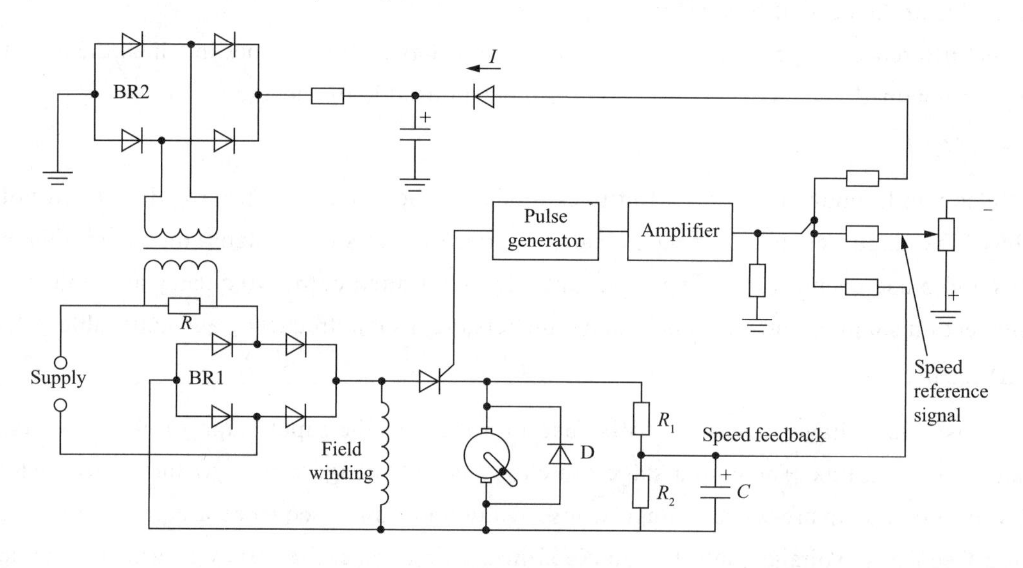

Figure 3.8 A single-quadrant thyristor converter system

Since the circuit is a single-quadrant converter, the speed of the motor shaft (which is the output from the system) can be controlled in one direction of rotation only. Moreover, regenerative braking cannot be applied to the motor; in this type of system, the motor armature can suddenly be brought to rest by dynamic braking (i.e. when the thyristor gate pulses are phased back to 180°, a resistor can be connected across the armature by a relay or some other means) .

Rectifier BR1 provides a constant voltage across the shunt field winding, giving a constant field flux. The armature current is controlled by a thyristor which is, in turn, controlled by the pulses applied to its gate. The armature speed increases as the pulses are phased forward (which reduces the delay angle of firing), and the armature speed reduces as the gate pulses are phased back.

The speed reference signal is derived from a manually operated potentiometer (shown at the right-hand side of Figure 3.8), and the feedback signal or output speed signal is derived from the resistor chain R_1 and R_2, which is connected across the armature. (Strictly speaking, the feedback signal in the system in Figure 3.8 is proportional to the armature voltage, which is proportional to the shaft speed only if the armature resistance drop, I_aR_a, is small.) Since the armature voltage is obtained from a thyristor, the voltage consists of a series of pulses; these pulses are smoothed by capacitor C. The speed reference signal is of the opposite polarity to the armature voltage signal to ensure that overall negative feedback is applied.

A feature of DC motor drives is that the load presented to the supply is a mixture of resistance, inductance, and back EMF diode D in Figure 3.8 ensures that the thyristor current commutates to zero when its anode potential falls below the potential of the upper armature connection, in the manner outlined before. In the drive shown in Figure 3.8, the potential of the thyristor cathode is equal to the back EMF of the motor while it is in a blocking state. Conduction can only take place during the time interval when the instantaneous supply voltage is greater than the back EMF. Inspection of Figure 3.9 shows that when the motor is running, the peak inverse voltage applied to the thyristor is much greater than the peak forward voltage. By connecting a diode in series with the thyristor, as shown, the reverse blocking capability of the circuit is increased to allow low-voltage thyristors to be used.

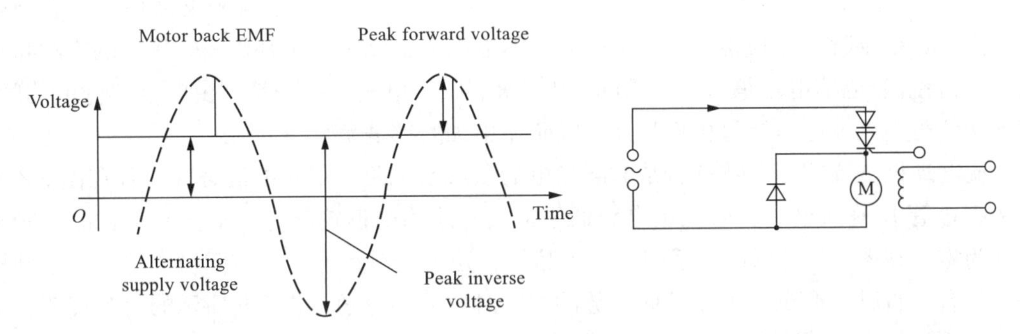

Figure 3.9 Illustrating the effect of motor back EMF on the peak inverse voltage applied to the thyristor

The waveform shown in Figure 3.10 is idealized waveform as much as it ignore the effects of armature inductance, commutator ripple, etc. Typical armature voltage waveform is shown in Figure 3.10. In this waveform the thyristor is triggered at point A, and conduction continues to point B when the supply voltage falls below the armature back EMF. The effect of armature inductance is to force the thyristor to continue to conduct until point C, when the fly-wheel diode prevents the armature voltage from reversing. When the inductive energy has dissipated (point D) , the armature current is zero and the voltage returns to its normal level, the transients having settled out by point E. The undulations on the waveform between E and F are due to commutator ripple.

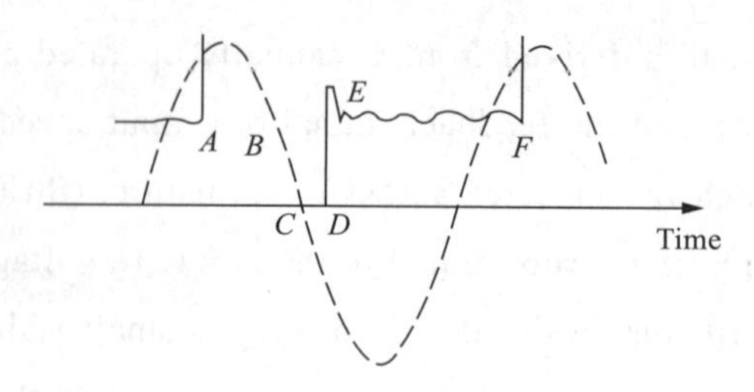

Figure 3.10 Typical armature voltage waveform

使用晶闸管的单相速度控制系统

一个单相晶闸管逆变器系统如图 3.8 所示。读者应该先忽略整流器 BR2 和它的相关电路(包括交流回路中的电阻 R)，因为这部分只有在具有保护功能时才需要，这将在下一节介绍。

因为该电路是一个单相转换器，只能在一个旋转方向控制电动机轴（系统的输出）的速度。而且，回馈制动不能用于电动机；在这种系统类型中，电动机电枢可以通过电气制动静止（例如，当晶闸管门极脉冲反相时，电阻可通过一个继电器或其他装置连接到电枢上)。

整流器 BR1 给并联励磁绕组提供一个稳定电压，产生稳定的磁通。电枢电流由一个晶闸管控制，该晶闸管又由加在它门极上的脉冲控制。脉冲正相时（减小启动延时角)，电枢转速增加；脉冲反相时，电枢转速减小。

速度参考信号可从人工操作的电位器（见图 3.8 右侧）上获得。反馈信号或输出转速信号可从连接在电枢上的电阻链 R_1 和 R_2 上获得。(严格地讲，图 3.8 系统中反馈信号只有当电枢电阻压降 I_aR_a 很小时，才与轴转速成正比的电枢电压成正比。用于补偿 I_aR_a 压降的方法将在阅读材料中讨论。) 电枢电压是从一个晶闸管上获得的，该电压包括一系列由电容 C 滤波的脉冲。速度参考信号与电枢电压信号极性相反，以确保施加的都是负反馈。

直流电动机装置的一个特征就是需要供电的负载是电阻、电导的混合，并且在图 3.8 中反电动势二极管 D 确保当晶闸管阳极电势低于前面叙述的电枢连接方式的上限时，晶闸管电流应换向为零。在图 3.8 所示拖动系统中，当晶闸管处于关断状态时，其阳极电势等于电动机反电势。只有在瞬时电源电压大于反向电势的间隔时它才会导通。图 3.9 所示的检测表明，电动机运行时晶闸管上峰值反向电压大于峰值正向电压。如图 3.9 所示，在晶闸管上串联一个二极管，电路的反向关断能力就会增强，所以允许使用低压晶闸管。

图 3.9 所示的波形是理想波形，因为忽略了电枢电感、换向器纹波等因素的影响。典型的电枢电压波形如图 3.10 所示。在该波形中，晶闸管在 A 点触发，一直到 B 点电源电压低于电枢反电势时导通。电枢电感的作用是使晶闸管保持到 C 点飞轮二极管使电枢电压反向之前导通。当电感能量消失（D 点)，电枢电流为零，电压恢复到它的正常水平，这个暂态过程最后稳定在 E 点。点 E、F 之前的纹波是由换向器引起的。

Self-Test

Translate the following phrases into English.

(1) 转换器。

(2) 自整流转换器。
(3) 整流器。
(4) 逆变器。
(5) 三相异步交流电动机。
(6) 步进电动机。
(7) 伺服电动机。
(8) 单相交流电动机。

New Words and Phrases

regulator *n.* 调整者，校准者，调整器，标准仪
steady-state *adj.* 不变的，永恒的
constant *n.* 常（系）数，恒（常）量，恒定（不变）值
adj. 不变的；恒（稳，固，坚）定的
regenerative *adj.* 再（新）生的，更新的，恢复的；回热（式）的；（正）反馈的，回授的
eddy current 涡流
dynamic *adj.* 动力（学）的，动（态）的，（不断）变化的，电动的，冲击的；有力的，高效能的，生动的
n. （原）动力，动态
reverse current 反向（反转）电流
thyristor *n.* 晶闸管
inverter *n.* 变换器，倒相器；倒换器；反演器；逆变器；变换电路，转换开关；反相旋转换流器；[计]“非”门（电路），变极器
rotor *n.* （电机的）转子
hoist *n.* 升起；吊起；推起；起重机
rectifier *n.* 整流器
pulse *n.* 脉冲
commutator *n.* 换向器，整流器
ripple *n.* 波纹
v. 起波纹
conduction *n.* 引流；输送，传播；传导，导电；传导性（率）；导热性（率）
transient *adj.* 短暂的，瞬时的
n. 瞬时现象
review *vt.* 回顾，复习
n. 回顾，复习，评论
concept *n.* 观念，概念
squirrel *n.* 松鼠
v. 储藏

induction *n.* 引入（导），诱导（作用）；感应（现象），电感，磁感；归纳（法，推理）
link *n.* 链环，连结物，链接
vt. 连结，联合
vi. 连接起来
generalize *vt.* 归纳，概括，推广，普及
semiconductor *n.* 半导体
transistor *n.* 晶体管
capacity *n.* 容量
capability *n.* （实际）能力，性能，容量，接受力
modulation *n.* 调节；调谐，调制
arrangement *n.* 排列，安排；协调；调解
solid-state *adj.* 固态的
terminology *n.* （-gies） 术语；专门名词
parameter *n.* 参数（量，项，词）
timing sequence 时序
comparison *n.* 比较，对照，比喻，比较关系
sine wave （三角的）正弦波
chopper *n.* 断路器，限制器，斩波器，交流变换器

Unit ④

Electrical Control System

Several types of control devices are used in industry to satisfy the control systems needs.

1. Switch

A switch is a device that either opens or closes a circuit. Although there are numerous types and styles of switches, they can be classified into the following categories, as shown in Figure 4.1.

(1) Locking and nonlocking.

(2) Normally opened and normally closed.

(3) Single throw and multiple throw.

(4) Single pole and multiple pole.

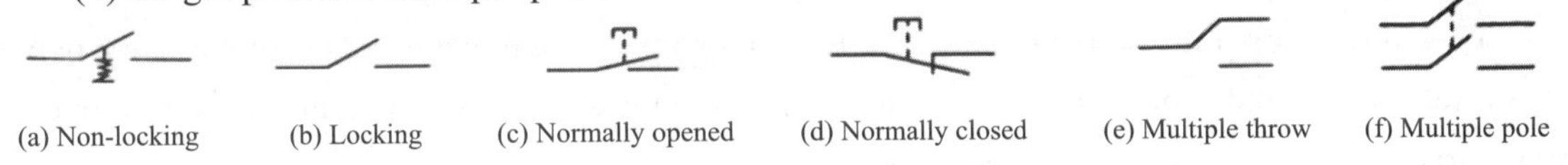

(a) Non-locking (b) Locking (c) Normally opened (d) Normally closed (e) Multiple throw (f) Multiple pole

Figure 4.1 Types of switches

Figure 4.2 shows various switches used in electrical control systems.

(a) Selector switch (b) Push-button switch (c) Limit switch

(d) Cross-switch (e) Thumbwheel switch (f) Slide switch

Figure 4.2 Various switches used in electrical control systems

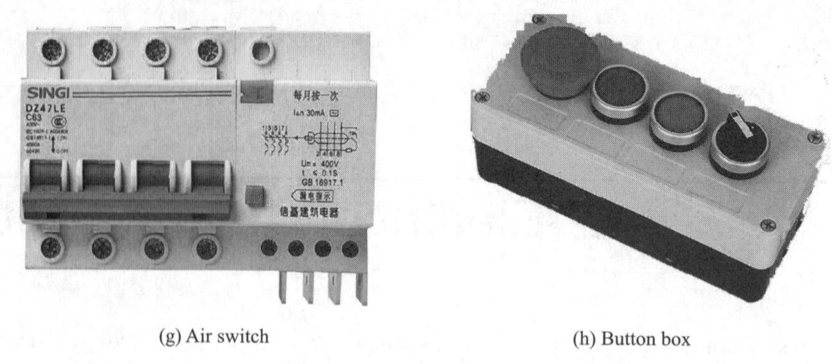

(g) Air switch (h) Button box

Figure 4.2 Various switches used in electrical control systems (continued)

2. Relay

A switch whose operation is activated by an electromagnet is called a relay (as shown in Figure 4.3). The contact and symbology for relays is usually the same as switch. Usually, the magnet is rated between 3 to 100 volts and a few hundred milliamps. Therefore, it is operated at very low power (current and voltage). A circuit carrying a much heavier rating can be switched using a relay, however, the two circuits are totally separated.

When a relay operates, the contacts dot all open or lose instantaneously. There may be a delay of several milliseconds between the operation of two contacts of the same relay. In the design of a relay circuit, this delay must always be taken into account.

Based on the discussion, one can see that a relay a magnet-operated contact switch. The contact switch inside a relay also can be classified by the number of poles and throws. Although most relays are single throw, it is very common to have multiple-pole relay.

Figure 4.3 Relay

3. Other types of low-voltage electrical equipment (see Figure 4.4)

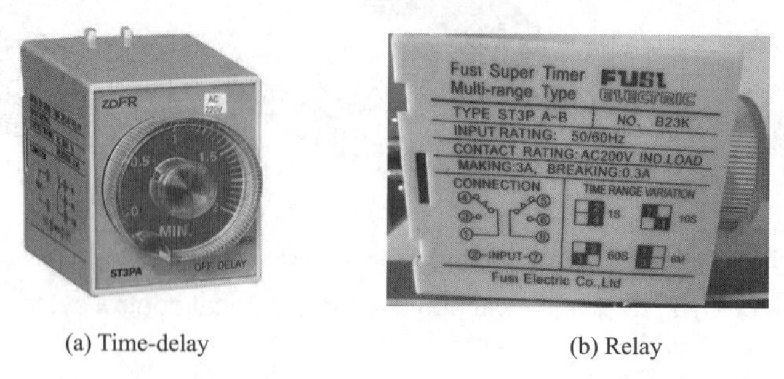

(a) Time-delay (b) Relay

Figure 4.4 Other types of low-voltage electrical equipment

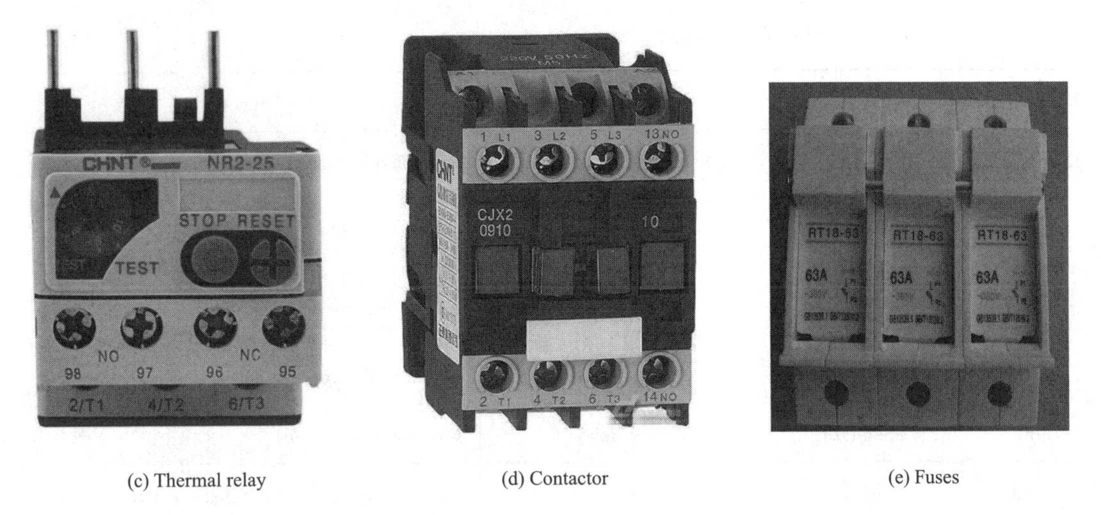

(c) Thermal relay (d) Contactor (e) Fuses

Figure 4.4 Other types of low-voltage electrical equipment (continued)

4. Electrical control system design

Electrical circuit diagram is shown as Figure 4.5. The component of electrical circuit diagram is shown in Table 4.1.

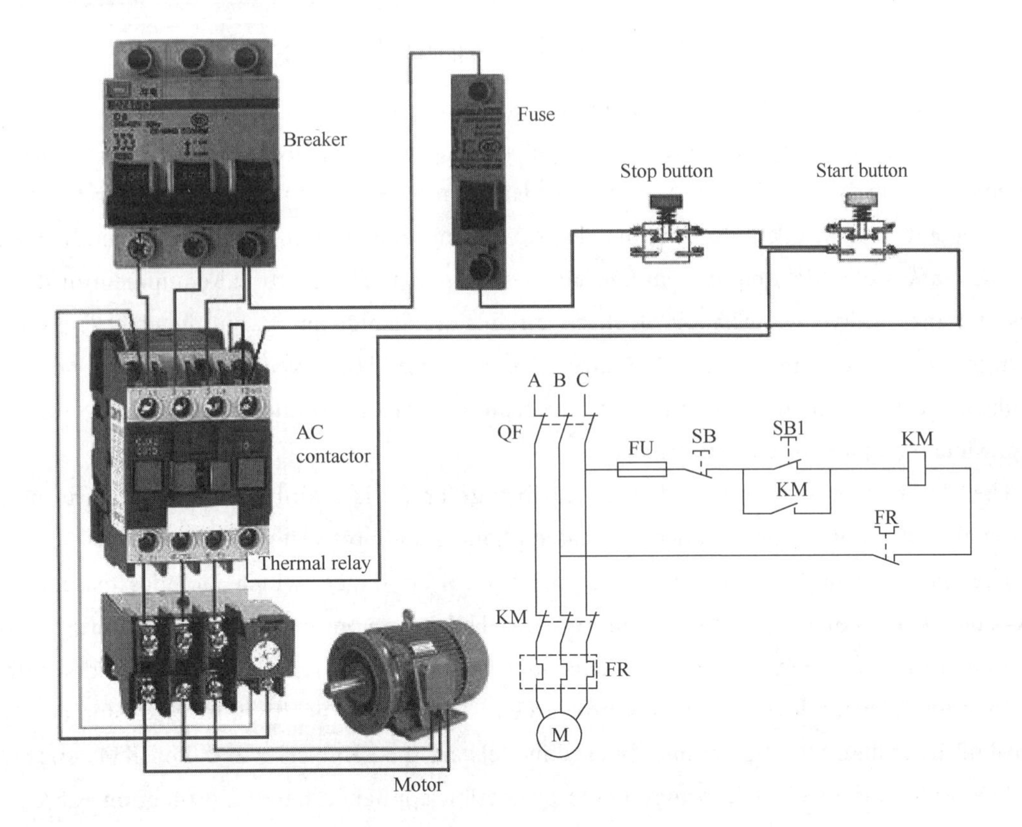

Figure 4.5 Electrical circuit diagram

Table 4.1 The component of electrical circuit diagram

Letter symbol	Chinese name	English name	Function
QS	空气开关	automatic air-break switch	Turn on or off the circuit
FU	熔断器	fuse	Short-circuit protection
KM	接触器	contactor	Turn on or off the circuit frequently
KT	时间继电器	time-delay relay	To turn off or turn on the circuit of the output signal after a predetermined time
FR	热继电器	thermal relay	Overload protection
SB	按钮	button	For operating a given command signal
SQ	行程开关	travel switch	Limit control

Extension Materials

Crane's electrical control system

The crane steel structure is responsible for the load support; the crane motor is responsible for its movement. Crane body movements, such as starting, running, reversing and stop all are done by electrical or hydraulic control system. In order to crane movements smooth, accurate, safe and reliable operation is inseparable from the electric transmission, control and protection effectively.

The crane's electrical equipment and electrical circuits. The electrical equipment of different types of cranes is diverse, and their electrical circuits are also different, but the electric circuits of the largely consists of main circuit, control circuit, protection circuit, etc. There is no need to introduce the main electrical equipment and basic electrical circuits of the typical GM bridge crane, which is typical of electric cranes.

The electrical equipment of the general bridge crane is mainly used for electric motors, braking electromagnets, controlling electrical appliances and protecting electricity device.

The operating appliance is also called the control appliance, which includes the controller, the contactor, the control screen and the resistor. The main controller is used for large capacity motors or heavy, frequently starting events such as garb operations. It usually works with the corresponding contact device in the control screen to realize the positive, reverse, stop and speed control of the main motor. Its commonly used models are the LK4 series and the LKI4 series.

The protection of electric bridge crane protection appliance has the protection cabinet, the control screen, the current relay, the travel limit of each institution position, emergency switch, various safety interlock switch and fuse, etc. Be sensitive to the protection of electrical appliances, ensure the safety and reliability of the work, and ensure the safety of crane.

起重机电气控制系统

起重机钢结构负责负载支撑；起重电动机负责起重机的运动；起重机机体的启动、运行、换向和停止等动作均由电气或液压控制系统完成，从而使起重机动作平稳、准确、安全可靠，这与电气传动、控制和保护密不可分。

起重机的电气设备和电路：不同类型的起重机的电气设备是多种多样的，其电气回路也不一样，但是电气电路主要由主电路、控制电路、保护电路等组成。此处不需要介绍典型的 GM 桥式起重机的主要电气设备和基本电路，这是典型的电动起重机。

通用桥式起重机的电气设备主要用于电动机、制动电磁铁、电气控制和电气设备保护。

操作装置又称控制装置，包括控制器、接触器、控制屏和电阻器。主控制器用于大容量电动机或大型、经常启动的操作，如抓取操作。它通常与控制屏上相应的触点装置一起工作，以实现主电动机的正、反、停和速度控制。它常用的型号是 LK4 系列和 LKI4 系列。

电桥起重机保护装置有保护柜、控制屏、电流继电器、各机构位置的行程限制、应急开关、各种安全联锁开关和熔丝等。对电气设备的保护敏感，可确保工作的安全可靠，保证起重机的安全。

Self-Test

(1) Translate the following phrases into Chinese.

① cross-switch.

② thumbwheel switch.

③ slide switch.

④ thermal relay.

⑤ AC contactor.

⑥ travel switch.

(2) Translate the following sentences.

① The main switches, fuses and various meters are to be examined and to be renewed if found out of order.

② The voltmeters are to be calibrated.

③ The circuit-breaker fails to close or trip.

④ After No.1 circuit breaker has been refitted, test the following item:

A: No-voltage protection.

B: reverse power protection.

⑤ Treat the contact members with great care. If contacts are badly burned or pitted, polish them with a dry cloth or sand paper.

⑥ The relay does not function properly, due to bent shaft (improper clearance, loose or broken lead wire).

⑦ The relay does not function at all, due to loose connection of wire (breaking of wire, improper setting).

⑧ Carbon brushes should be replaced by the ones of the same specification.

(3) According to the electrical diagram shown in Figure 4.5, write out the instructions for the control system.

New Words and Phrases

electrical control 电气控制
relay-contactor 继电器－接触器
digital control technology 数字控制技术
control object 控制对象
typical *adj.* 典型的；特有的，特别的；代表性的
apparatus *n.* 仪器，器械；机器；机构
low-voltage electrical apparatus 低压电气装置
(QF) automatic air-break switch 自动开关
(FU) fuse 熔断器
(KM) contactor *n.* 接触器；触点
relay *n.* 继电器
vt. 转播，传达；使接替；分程传递
(SB) push button *n.* 按钮
overtravel-limit switch (travel switch) 行程开关
(KV) voltage relay (zero-sequence voltage relay) 电压继电器
(KI) current relay 电流继电器
(KT) time relay 时间继电器
coil (winding) *n.* （一）卷，（一）盘；盘卷之物；线圈
normally open contact 常开触点
normally closed contact 常闭触点
conformation principle 构造原理
electrical control system 电气控制系统
control requirement 控制要求
electrical schematic diagram 电气原理图
element symbol 元件符号
time control 时间控制
interlocked control 互锁控制，联锁控制
thermal relay 热继电器
overload protection 过载保护，过负荷保护

CHAPTER 3

Industrial Control System

Objectives:

(1) Describe the English name for various types of control devices.
(2) Recognize the nameplate of various control devices.
(3) Know the PLC control system.
(4) Know the pneumatic control system.
(5) Know the sensing devices.

Unit ⑤

Control System

1. Some examples used in automatic systems (see Figure 5.1)

(a) Logistics sorting automatic line

(b) Automatic production line of mobile phone shell

(c) Automobile automatic production line

(d) Automatic production line for packing moon cake

Figure 5.1 Application of automatic control system

2. Some control devices used in automatic system

Several types of control devices are used in industry to satisfy the following control needs (see Figure 5.2-Figure 5.6).

Figure 5.2 Mechanical control

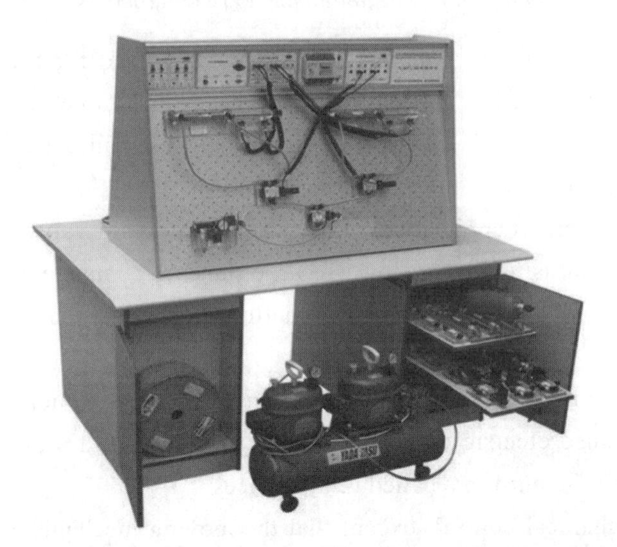

Figure 5.3 Pneumatic control

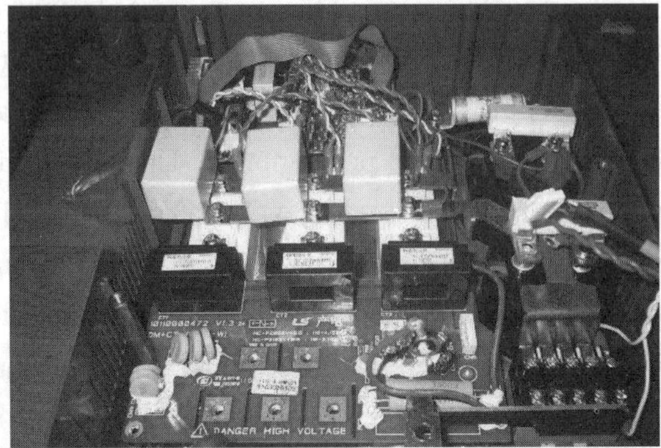

Figure 5.4 Electronic control

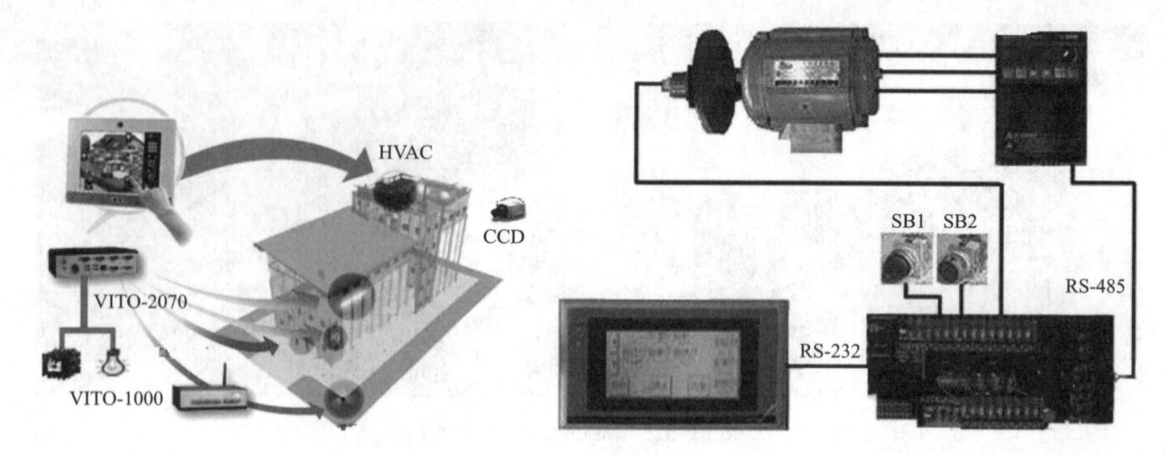

Figure 5.5 Computer control

Figure 5.6 Programmable logic control (PLC)

Mechanical control includes cams and governors. Although they have been used for the control of very complex machines, to be cost effectively, today they are used for simple and fixed-cycle task control. Some automated machines, such as screw machines, still use cam-based control. Mechanical control is difficult to machines and is subject to wear.

Pneumatic control is still very popular for certain applications. It uses compressed air, valves, and switches to construct simple control logic, but is relatively slow. Because standard components are used to construct the logic, it is easier to build than a mechanical control. Pneumatic control parts are subject to wear.

As does a mechanical control, an electromechanical control uses switches, relays, timers, counter, and so on, to construct logic. Because electric current is used, it is faster and more flexible. The controllers using electromechanical control are called relay devices.

Electric control is similar to electromechanical control, except that the moving mechanical components in an electromechanical control device are replaced by electronic switches, which works faster and is more reliable.

Computer control is the most versatile control system. The logic of the control is programmed into the computer memory using software. It not only can be for machine and manufacturing system control, but also for data communication. Very complex strategies with extensive computations can be programmed. The first is the interface with the outside world. Internally, the computer uses a low voltage (5-12 Volts) and a low current (several milliamps). Machine requires much higher voltages (24 V, 110 V, or 220 V) and current (measured in amps). The interface not only has to convert the voltage difference, but also must filter out the electric noise usually found in the shop. The interface thus must be custom-built for each application.

A relay device is the main components of PLC and consists of the front display panel with switches, relays, timers, and counters.

3. The terminology to describe control systems

Automatic control has played a vital role in advancement of engineering and science. In addition to its extreme importance in space-vehicle, missile-guidance, and aircraft-piloting systems, etc. automatic control has become an importance and integral part of modern manufacturing and industrial processes. For example, automatic control is essential in such industrial operations as controlling pressure, temperature, humidity, viscosity, and flow in the process industries, tooling, handing, and assembling mechanical parts in the manufacturing industries, among many others.

Since advance in the theory and practice of automatic control provide means for attaining optimal performance of dynamic systems, improve the quality and lower the cost of production, expand the production rate, relieve the drudgery of many routing, repetitive manual operations, etc. Most engineers and scientists must now have a good understanding of this field.

Applications of modern control theory to such nonengineering fields as biology, economics, medicine, and sociology are now under way, and interesting and significant results can be expected in the near future.

Next we shall introduce the terminology necessary to describe control system.

1) Plants

A plant is a piece of equipment, perhaps just a set of machine parts functioning together, the purpose of which is to perform a particular operation. Here we shall call any physical object to be controlled (such as a heating furnace, a chemical reactor, or a spacecraft) a plant.

2) Process

The Merriam-Webster Dictionary defines a process to be a natural, progressively continuing operation or development marked by a series of gradual changes that succeed one another in a relatively fixed way and lead toward a particular result or end; or an artificial or voluntary, progressively continuing operation that consists of a series of controlled actions or movements systematically directed toward a particular result or end.

3) System

A system is a combination of components that act together and perform a certain objective. A system is not limited to abstract, dynamic phenomena such as those encountered in economics.

4) Feedback control system

A feedback control system is one which tends to maintain a prescribed relationship between the output and the reference input by comparing these and using the difference as a means of control.

5) Servomechanisms

A servomechanism is a feedback control system in which the output is some mechanical position, velocity, or acceleration. Therefore, the terms servomechanism and position (or velocity or acceleration) control system are synonymous. Servomechanisms are extensively used in modern industry.

6) Automatic regulating systems

An automatic regulating system is a feedback control system in which the reference input or the desired output is either constant or slowly varying with time and in which the primary task is to maintain the actual output at the desired value in the presence of disturbances.

A home heating system in which a thermostat is the controller is an example of an automatic regulating system. In this system, the thermostat setting (the desired temperature) is compared with the actual room temperature. A change in the desired room temperature is a disturbance in this system. The objective is to maintain the desired room temperature despite changes in outdoor temperature.

7) Process control systems

An automatic regulating system in which the output is a variable such as temperature, pressure, flow, liquid level, or pH is called a process control system. Process control is widely applied in industry. For example, a preset program may be such that the furnace temperature is raised to given temperature in a give time interval and then lowered to another given temperature in some other given time interval. In such program control the set point is varied according to the preset time schedule.

Extension Materials

General use of microcomputer control

The requirements on a process-control computer were neatly matched with progress in integrated-circuit technology. The development of minicomputer technology combined with the increasing knowledge gained about process control with computers caused a rapid increase in applications of computer control. Special process-control computers were announced by several manufacturers.

As an example of a single-loop controller for process control, such systems were traditionally implemented using pneumatic or electronic techniques, but they are now always computer-based. The controller has the traditional proportional, integral, and derivative actions, which are implemented in a microprocessor. In this particular case, the regulator is provided with automatic tuning, gain scheduling, and continuous adaptation of feedforward and feedback gains. These functions are difficult to implement with analog techniques.

The next development phase in industrial process-control systems was facilitated by the emergence of common standard in computing, making it possible to integrate virtually all computers and computer systems in industrial plants into a monolithic whole to achieve real-time exchange of data across what used to be closed system borders. All database instances are addressed by their symbolic names for ease of access. The functional relationships among database instances are defined by application programs that are developed graphically by laying out the required function blocks on diagram sheets, the connecting these blocks to other blocks so as to process inputs and outputs. These are function blocks as simple as logic AND and OR gates and as complex as adaptive controllers. Interconnections can be given symbolic names for easy reference and the programming tool assigns page references automatically. The result is in the form of graphic program diagrams to IEC's standards to the applicable extent that diagrams can be easily understood by all concerned. Process operators then monitor and control the process from a variety of live displays, from process schematics and trend graphs to alarm lists and reports. Live measured values and status indications reveal the current situation.

计算机控制的普遍应用

对过程控制计算机的要求与集成电路技术的发展密切相关。小型计算机技术的发展以及人们对过程控制与计算机知识的日益了解使计算机控制的应用日益广泛。许多厂家已经宣称能够制造专用过程控制计算机。

例如单一回环过程控制的控制器，过去使用气动或者电子技术，但现在通常使用计算机。控制器传统的比例、积分和微分功能由微处理器实现。在这种情况下，控制器可以自动调谐、获得时序并持续进行前馈增益和反馈调节。这些功能对于模拟技术来说很难实现。

计算机通用标准使工业过程控制系统进入了一个新的发展阶段，使之有可能把工业设备中的所有计算机和计算机系统集成为一个实际的整体，实现各个闭式系统之间数据的实时交流。为了便于访问，所有的数据库实例的地址就是它们的符号名。各个数据实例之间和功能关系由应用程序确定，把需要的功能模块画成框图，然后连接这些模块以便处理输入和输出。这些功能模块有的十分简单，例如逻辑“与”门和“或”门；有的比较复杂，例如自适应控制装置等。它们的相互连接方式用符号标识，编程工具自动给出每页的说明。结果用国际电工委员会标准的图形编程形式给出，非常容易理解和应用。操作者从各种各样的实况显示、过程图表、趋势图以及故障警示表格和报告监测并控制整个过程。现场测量的数据和状态标识揭示了当前的状态。

Self-Test

(1) Translate the following phrases into Chinese.

① servomechanism.

② automatic regulating system.

③ feedback control system.

④ mechanical control.

⑤ pneumatic control.

⑥ electric control.

(2) Identify the following to be True or False.

① For example, automatic control is essential in tooling, handling; and assembling mechanical parts in the manufacturing industries, among many others.( )

② Advances in the theory and practice of automatic control provide ways for attaining optimal performance of dynamic systems improve the quality and lower the cost of production, expand the production rate, etc.( )

③ Applications of modern control theory to such nonengineering fields as biology, economics, medicine, and sociology are currently in progress.( )

④ In process control system, the set point is changing according to the preset time schedule.( )

New Words and Phrases

humidity *n.* 湿气，潮湿，湿度
tooling *n.* 用刀具加工，工具，机床安装
servomechanism *n.* 伺服机构，自动控制装置
terminology *n.* 术语
plant *n.* 工厂，对象，车间，设备
thermostat *n.* 自动调温器，温度调节装置
preset *vt.* 事先调整，预先安置，预先调试
set point 设定点，设定值，凝结点，调定点

Unit 6

Fundamentals and Application of Sensors

1. Some applications of sensors in our daily life (see Figure 6.1)

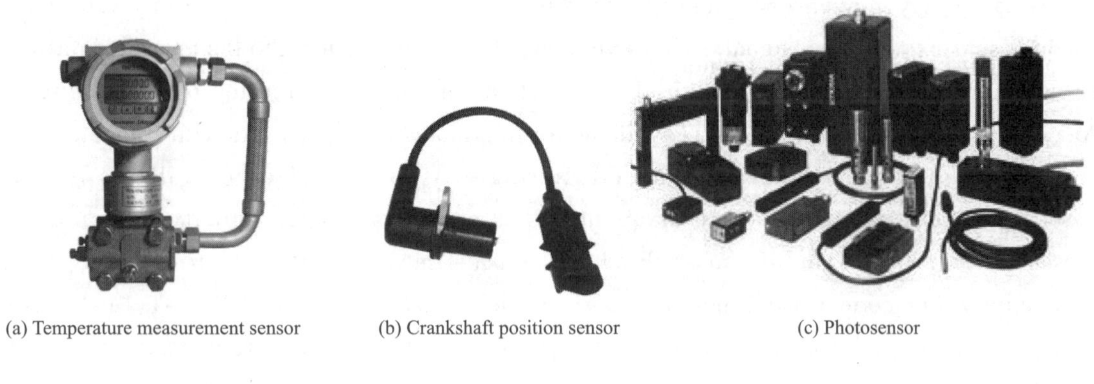

(a) Temperature measurement sensor (b) Crankshaft position sensor (c) Photosensor

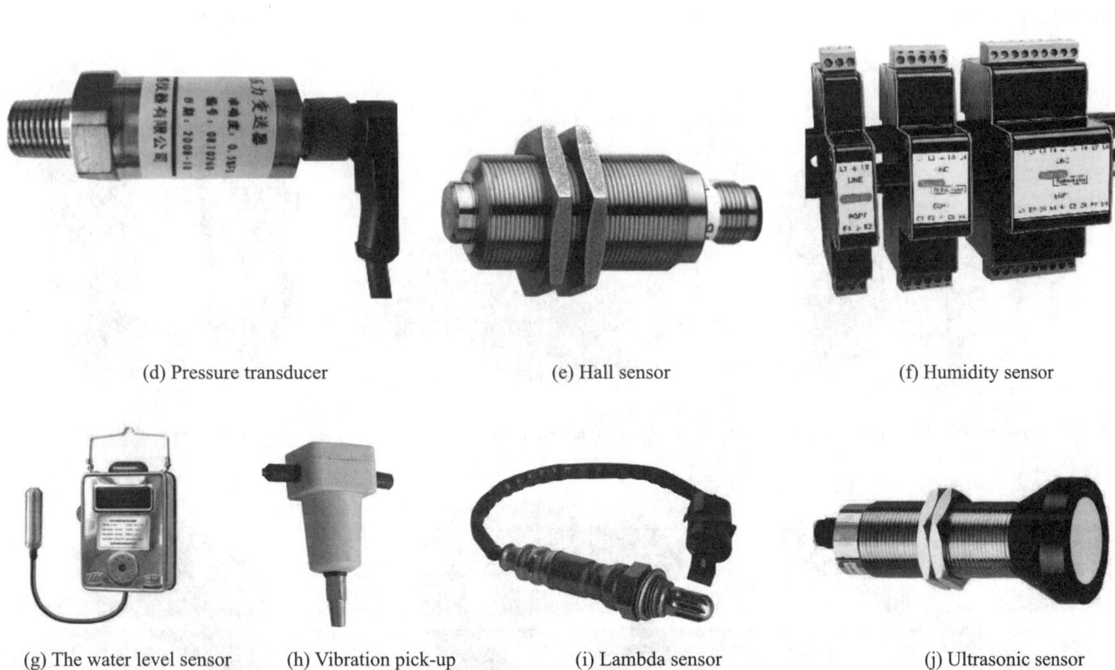

(d) Pressure transducer (e) Hall sensor (f) Humidity sensor

(g) The water level sensor (h) Vibration pick-up (i) Lambda sensor (j) Ultrasonic sensor

Figure 6.1 Various types of sensors

2. Something about sensor technology

A sensor is a device which produces a signal in response to its detecting or measuring a property, such as position, force, torque, pressure, temperature, humidity, speed, acceleration, or vibration. Traditionally, sensors (such as actuators and switches) have been used to set limits on the performance of machines.

Common examples are:

(1) Stops machine tools to restrict work table movements;

(2) Pressure and temperature gages with automatics shut-off features;

(3) Governors on engines to prevent excessive speed of operation.

Sensor technology has become an important aspect of manufacturing processes and system. It is essential for proper data acquisition and for the monitoring, communication, and computer control of machines and systems.

Because they convert one quantity to another, sensors often are referred to as transducers. Analogy sensors produce a signal, such as voltage, which is proportional to the measured quantity. Digital sensors have numeric or digital outputs that can be transferred to computers directly. Analog-to-digital-converter (ADC) is available for interfacing analog sensors with computers.

Sensors that are of interest in manufacturing may be classified generally as follows: Machanical sensors measure such as quantities as positions, shape, velocity, force, torque, pressure, vibration, strain, and mass. Electrical sensors measure voltage, current, charge, and conductivity. Magnetic sensors measure magnetic field, flux, and permeability. Thermal sensors measure temperature, flux, conductivity, and special heat. Other types are acoustic, ultrasonic, chemical, optical, radiation, laser, and fiber-optic. Figure 6.2 shows various sensors.

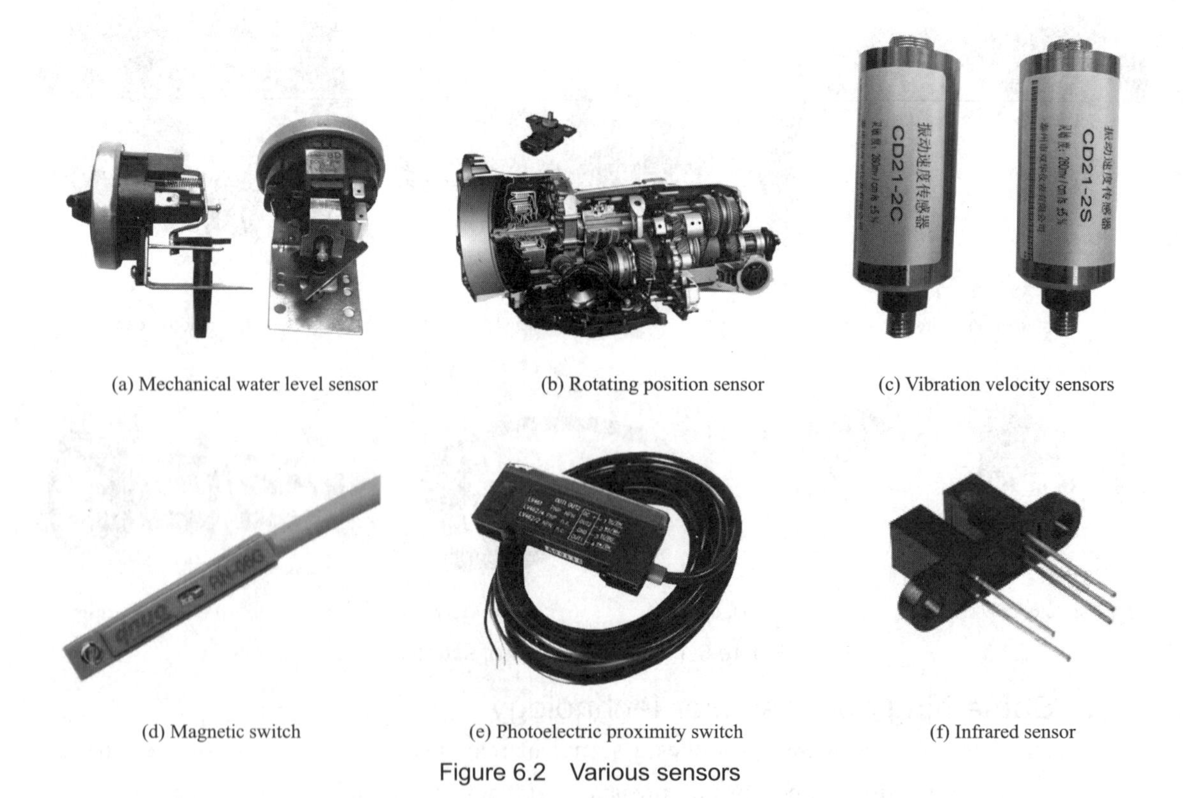

(a) Mechanical water level sensor (b) Rotating position sensor (c) Vibration velocity sensors

(d) Magnetic switch (e) Photoelectric proximity switch (f) Infrared sensor

Figure 6.2 Various sensors

(g) Hall magnetic sensor (h) Laser side distance sensor (i) Laser displacement sensor

(j) Clamp ammeter (k) Acoustic sensor (l) Rotary encoders

(m) Heat flow sensor (n) Temperature sensors

Figure 6.2 Various sensors (continued)

Depending on its application, a sensor may consist of metallic, nonmetallic, organic, or inorganic materials, as well as fluids, gases, plasmas, or semiconductors. Using the special characteristics of these materials, sensors covert the quantity or property measured to analog or digital output. The operation of an ordinary mercury thermometer, for example, is based on the difference between the thermal expansion of mercury and that of glass.

Similarly, a machine part, a physical obstruction, or barrier in a space can be detected by breaking the beam of light when sensed by a photoelectric cell. A proximity sensor (which senses and measures the distance between it and an object or a moving member of a machine) can be based on acoustics, magnetism, capacitance, or optic. Other actuators contact the object and take appropriate action (usually by electromechanical means). Sensors are essential to the conduct of intelligent robots, and are being developed with capabilities that resemble those of humans (smart sensors see the following).

Tactile sensing is the continuous of variable contact forces, commonly by an array of sensors. Such a system is capable of performing with an arbitrary three-dimensional space.

In visual sensing (machine vision, computer vision), camera optically sense the presence and shape of the object. A microprocessor then processes the image (usually in less than one second), the image is measured, and the measurements are digitized (image recognition). Machine vision is suitable particularly for inaccessible part, in hostile manufacturing environments, for measuring a large number of small features, and in situations where physics contact with the part may cause damage.

Small sensors have the capability to perform a logic function, to conduct two-way communication, and to make a decision and take appropriate actions. The necessary input and the knowledge required to make a decision can be built into a smart sensor. For example, a computer chip with sensors can be programmed to turn a machine tool off when a cutting tool fails. Likewise, a smart sensor can stop a mobile robot or a robot arm from accidentally coming in contact with an object or people by using quantities such as distance, heat, and noise.

Sensor fusion basically involves the integration of multiple sensors in such a manner where the individual data from each of the sensors (such as force, vibration, temperature, and dimensions) are combined to provide a higher level of information and reliability. A common application of sensor fusion occurs when someone drinks a cup of hot coffee. Although we take such a quotidian event for granted, it readily can be seen that this process involves data input from the person's eyes, lip, tongue and hands. Through our basic senses of sight, hearing, smell, taste, and touch, there is real-time monitoring of relative movements, positions, and temperatures. Thus if the coffee is too hot, the hand movement of the cup toward the lip is controlled and adjusted accordingly.

The earliest applications of sensor fusion were in robot movement control, missile flight tracking, and similar military applications. Primarily because these activities involve movements that mimic human behavior. Another example of sensor fusion is a machine operation in which a set of different integrated sensors monitors: the dimensions and surface finish of workpiece; tool forces, vibrations and wear; the temperature in various regions of the tool-workpiece system, and the spindle power.

An important aspect in sensor fusion is sensor validation: the failure of one particular sensor is detected so that the control system maintains high reliability. For this application, the receiving of redundant data from different sensors is essential. It can be seen that the receiving, integrating of all data from various sensors can be a complex problem.

With advances in sensor size, quality, and technology and continued developments in computer-control systems, artificial neural networks, sensor fusion has become practical and available at low cost. Movement is relatively independent of the number of components, the equivalent of our body, waist is a rotary degree of freedom. We have to be able to hold his arm. Arm can be bent, then this three degrees of freedom. Meanwhile there is a wrist posture adjustment to the use of the three autonomy, the general robot has six degrees of freedom. We will be able to space the three locations, three postures, the robot fully achieved, and of course we have less than six degrees of freedom.

Fiber-optic sensors are being developed for gas-turbine engines. These sensors will be installed in critical locations and will monitor the conditions inside the engine, such as temperature, pressure, and flow of gas. Continuous monitoring of the signals from these sensors will help detect possible engine problems and also provide the necessary data for improving the efficiency of the engines.

Extension Materials

Application of biosensors

In biomedicine and biotechnology, sensors which detect analytes thanks to biological component, such as cells, protein, nucleic acid or biomimetic polymers, are called biosensors, as shown in Figure 6.3. Whereas a non-biological sensor, even organic (carbon chemistry), for biological analytes is referred to as sensor or nanosensor (such a microcantilevers). This terminology applies for both in vitro and in vivo applications. The encapsulation of the biological component in biosensors, presents with a slightly different problem that ordinary sensors, this can either be done by means of a semipermeable barrier, such as a dialysis membrane or a hydrogel, a 3D polymer matrix, which either physically constrains the sensing macromolecule or chemically. Figure 6.4 shows application of biosensors.

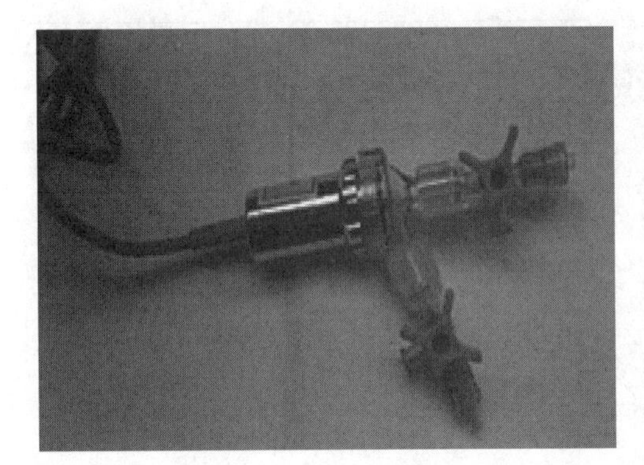

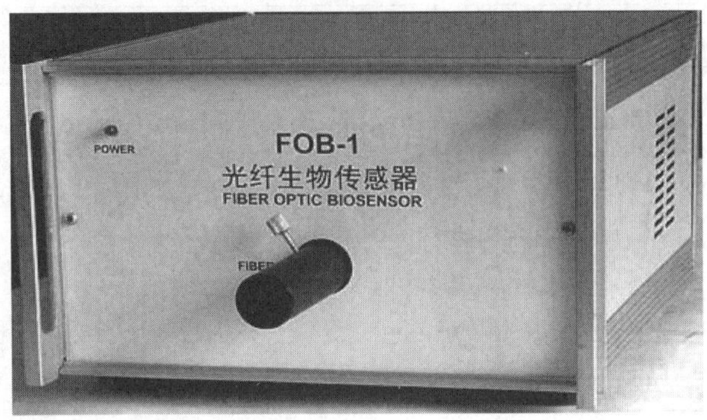

Figure 6.3 The biosensor

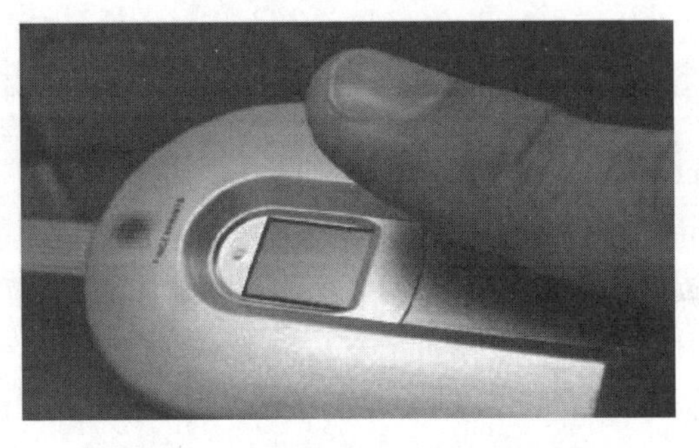

Figure 6.4 Application of biosensors

Figure 6.4 Application of biosensors (continued)

生物传感器的应用

在生物医学和生物技术中，由于生物成分（如细胞、蛋白质、核酸或仿生聚合物）而检测分析物的传感器称为生物传感器，如图 6.3 所示。而用于生物分析的非生物传感器，甚至是有机体(碳化学),称为传感器或纳米传感器(如微悬臂梁)。这一专业术语适用于试管的内外应用。生物传感器中生物成分的封装存在着一个与普通传感器略有不同的问题，即可以通过透析膜或水凝胶等半透性屏障来实现，这是一种在物理上或化学上限制传感大分子的三维聚合物基质。图 6.4 所示为生物传感器的应用。

Self-Test

Translate the following phrases into Chinese.

(1) biosensor.

(2) temperature sensor.

(3) laser side distance sensor.

(4) magnetic switch.

(5) lambda sensor.

(6) infrared sensor.

(7) acoustic sensor.

New Words and Phrases

temperature measurement sensor　水温测量传感器

crankshaft position sensor　曲轴位置传感器

photosensor　*n.*　光敏元件；光敏器件；光电探测器；光电传感器

pressure transducer　压力传感器

hall sensor　霍尔传感器

humidity sensor　湿度传感器
the water level sensor　水位传感器
vibration pick-up　振动传感器；拾振仪
lambda sensor　氧传感器
ultrasonic sensor　超声波传感器
biological effect　生物效应
infrared　*adj*.　红外线的
acoustic　*adj*.　声学的
transducer　*n*.　传感器

Unit 7

Inverter

1. The brand of the mainstream inverter in the market (see Figure 7.1)

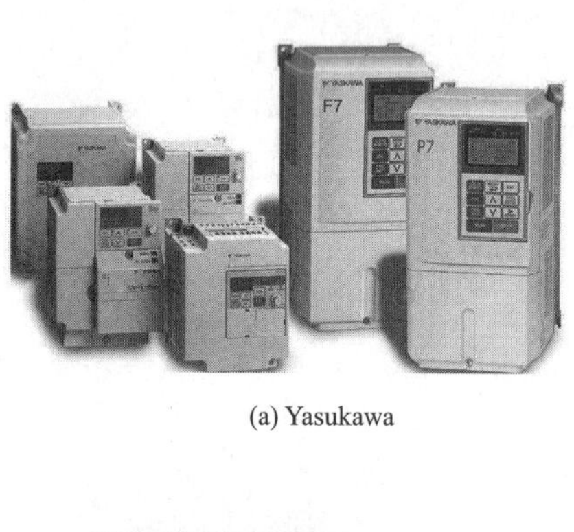

(a) Yasukawa

(b) Inovance

(c) Mitsubishi

(d) Siemens

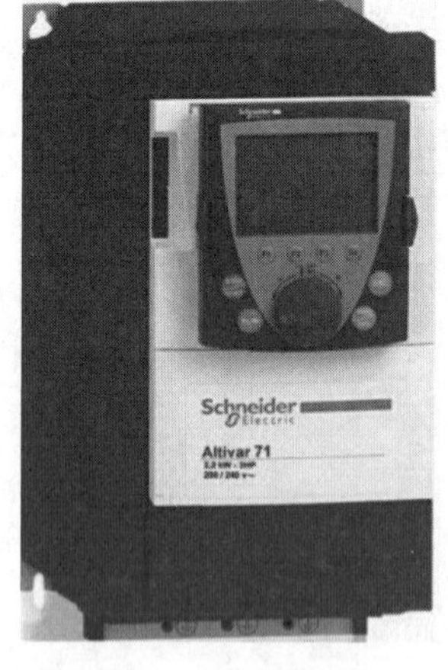

(e) Schneider

(f) Omron

Figure 7.1 The brand of the mainstream inverter in the market

2. The principle of frequency inverter

The converter principle knowledge is the basic principle of the converter. The frequency converter foundation VVVF is the abbreviation for variable voltage and variable frequency, meaning to change the voltage and change frequency, which is what people refer to as variable-voltage frequency. CVCF is the abbreviation for constant voltage and constant frequency. We use the power of the divided into AC power and DC power supply, DC power supply are mostly general by AC power through a transformer, rectifier after filtering. AC power supply is about 95 percent of the total power supply. Whether for home or used in factories, single-phase AC power supply and three-phase AC power supply, the voltage and frequency according to the provisions of the countries have certain standards, as specified in our country, direct user single phase alternating current (AC) is 220 V, three-phase AC voltage is 380 V, 50 Hz, power supply voltage and frequency of other countries may be in different voltage and frequency in our country, such as a single-phase 100 V/60 Hz, three-phase 200 V/60 Hz, etc., the standard of the voltage and frequency of the AC power supply AC power frequency. In general, the device that sets the voltage and frequency constant to the alternating current of alternating current to voltage or frequency is called a "frequency changer". In order to produce a variable voltage and frequency, the device first changes the alternating current of the power supply to direct current (DC), a process called rectifier. A device that transforms DC to AC, whose scientific term is inverter. The general inverter is the inverter power source that inverses DC power supply to certain fixed frequency and voltage. In a word, the drivers with adjustable voltage are called inverter.

3. The wiring and installation of Mitsubishi inverter

1) Basic wiring diagram (see Figure 7.2)

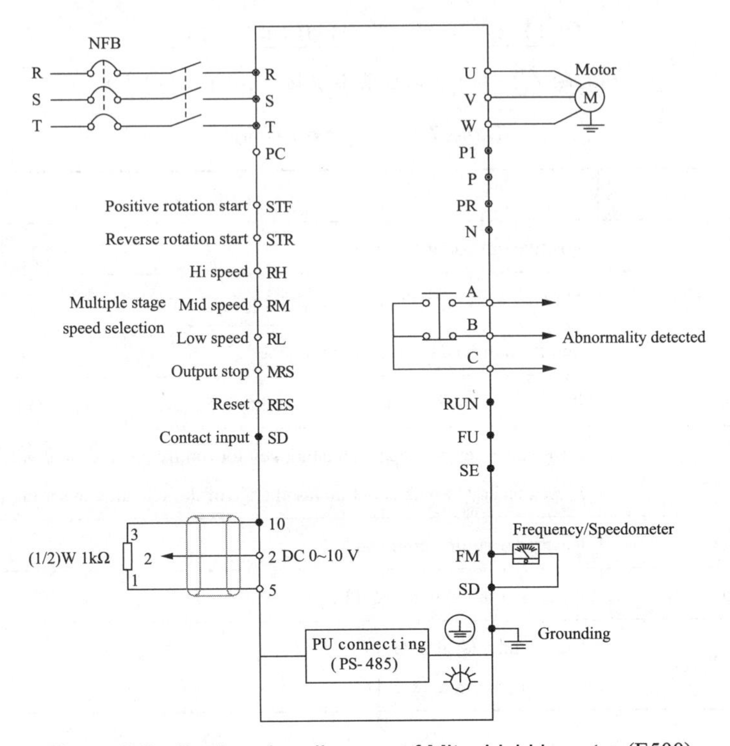

Figure 7.2 Basic wring diagram of Mitsubishi inverter (E500)

2) Digital keypad operation

Keypad of Mitsubishi inverter (E500) is shown in Figure 7.3. Table 7.1 Shows the key expressions.

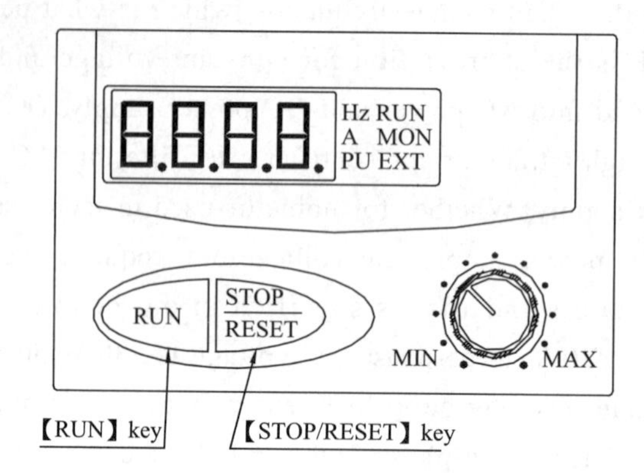

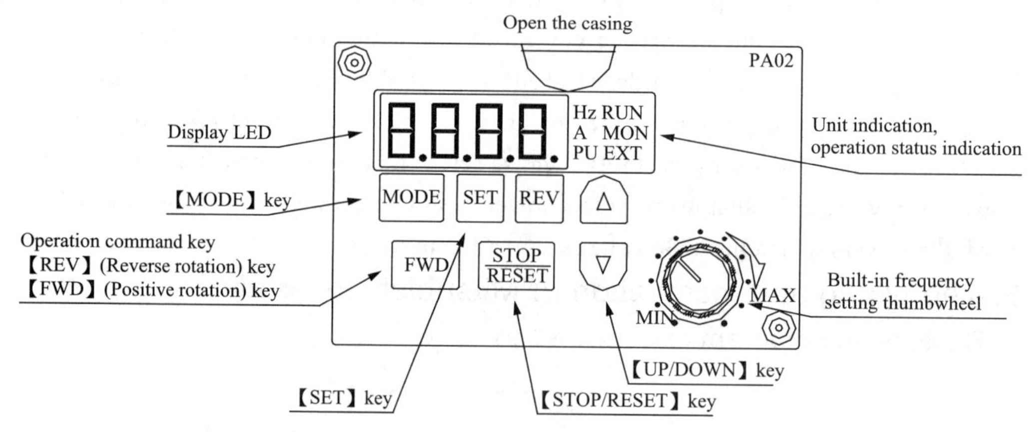

Figure 7.3 Keypad of Mitsubishi Inverter (E500)

Table 7.1 Key expression

Key	Expressions
【RUN】	Positive run command key
Built-in frequency setting device	May use analog method to change frequency setting
【MODE】 key	Selection of RUN mode, setting mode
【SET】 key	Please press the【SET】key after inputting frequency value, parameter set value
【△ / △】 key	Continue ascend/descend adjusting key for run frequency, valid only pressing. Press this key to make continuous change of the set value in setting mode
【REV】 key	Reverse rotation command key
【FWD】 key	Positive rotation command key
【STOP/RESET】 key	Run command stop key; Inverter abnormal reset key

Table 7.2 shows unit indication, run status indication.

Table 7.2 Unit indication, run status indication

Key	Contents
Hz	Light ON when the display value of the 7-stages monitor is frequency
A	Light ON when the display value of the 7-stages monitor is current
RUN	Light ON during the inverter running, constant light ON when positive rotation, blinking while reverse
MON	Light ON when under monitoring mode
PU	Light ON when under PU run mode
EXT	Light ON when under external run mode

4. The wiring and installation of Schneider inverter

Extracts from the programming manual of ATV31.

Figure 7.4 shows ATV31 variable speed drives for asynchronous motors. Figure 7.5 shows the function of each button.

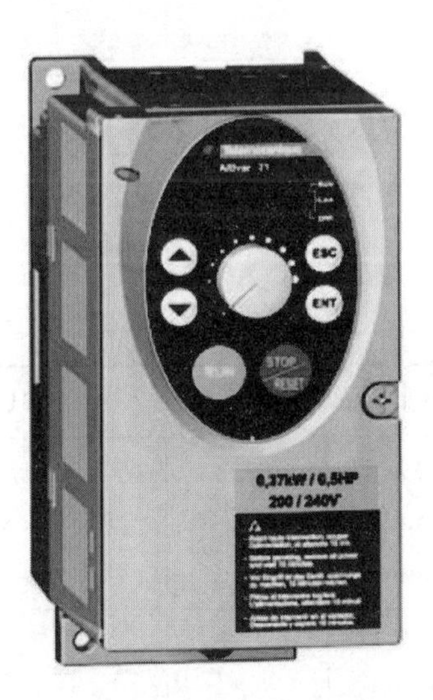

Figure 7.4 ATV31 variable speed drives for asynchronous motors

Four 7-segment displays
8.8.8.8
RUN
CAN
ERR
Two CAN open status LEDs
ESC
ENT
RUN
STOP RESET
Setting potentiometer
Stop/Reset button
Run/start button

Figure 7.5 The function of each button

1) Functions of the display and the keys

(ESC) Exits a menu (菜单) or parameter (参数), or aborts (清除) the displayed value to return to the previous value (数值) in the memory.

(ENT) Enters a menu or a parameter, or saves the displayed parameter or value.

(▲) Returns to the previous menu or parameter, or increases the displayed value.

(▼) Goes to the next menu or parameter, or decreases the displayed value.

Note:

(1) Pressing (▲) or (▼) does not store the selection.

(2) Press and hold down (>2 s) (▲) or (▼) to scroll through the data quickly.

(3) To save and store the displayed selection: ENT.

(4) The display flashes when a value is stored.

(5) Normal display, with no fault（故障）present and no starting:

① 43.0: Display of the parameter selected in the SUP menu [default selection（默认选项）: motor frequency].

② dCb: DC injection braking(in progress).

③ FSt: Fast stop.

④ nSt: Freewheel stop.

⑤ rdY: Drive ready.

⑥ tUn: Auto-tuning in progress.

(6) The display flashes to indicate the presence of a fault.

2) Install the drives

(1) Drive rating:

Single-phase supply voltage: 200-240 V 50 Hz/60 Hz.

Three-phase motor 200-240 V.

Table 7.3 shows the parameters.

Table 7.3 Parameters

Motor	Line supply(input)					Drive(output)			Altivar31
Power indicated on plate	Max. Line current		Max. prospective line current	Apparent power	Max. Inrush current	Nominal Current	Max. transient current	Power dissipation	Reference
	at 200 V	at 240 V							
kW/hp	A	A	kA	kV · A	A	A	A	W	
0.18/0.25	3.0	2.5	1	0.6	10	1.5	2.3	24	ATV31H018M2

(2) These power ratings and currents are given for an ambient temperature of 50 ℃ (122 ˚F) at the factory-set switching frequency, used in continuous operation (switching frequency factory setting 4 kHz for ATV31H 037M3. Above this factory setting, the drive will reduce the switching frequency automatically in the event of excessive temperature rise.

(3) Current on a line supply with the “Max. prospective line current” indicated and for a drive without any external options.

(4) Peak current on power-up for the max. voltage 240 × (1 ± 10%) V.

3) Mounting and temperature conditions

(1) Install the drive vertically at ± 10°.

(2) Do not place it close to heating elements.

(3) Leave sufficient free space to ensure that the air required for cooling purposes can circulate from the bottom to the top of the unit.

(4) Free space in front of the drive: 10 mm (0.39 in) minimum.

(5) When IP20 protection is adequate, it is recommended that the protective cover on the top of the drive is removed as shown in Figure 7.6.

(6) Removing the protective cover.

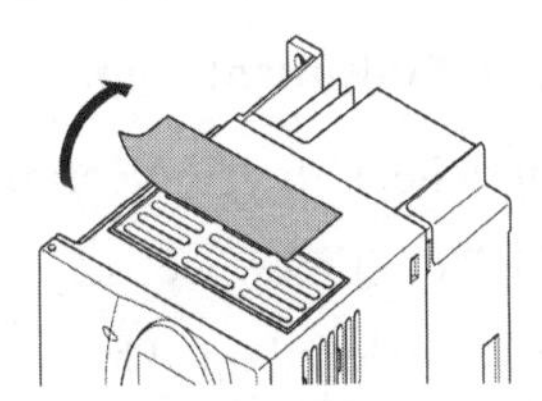

Figure 7.6 Removing the protective cover

4) Characteristics and functions of the control terminals (see Table 7.4)

Table 7.4 Characteristics and functions of the control terminals

COM	Analog I/O common	0 V
AL1	Analog voltage input	(1) Analog input: 0 V to +10 V (max. safe voltage 30 V). (2) Impedance: 30 kΩ. (3) Reaction time: (2 ± 0.5) ms. (4) Resolution: 10 bits; accuracy: ± 0.6%; linearity: ± 0.15% of max. value
10 V	power supply for reference potentiometer	(1) +10 V_c [(10.5 ± 0.5) V]. (2) 10 mA max.
AL2	Analog voltage input	(1) The bipolar analog input: 0 V to +10 V_c (max. safe voltage 24 V). (2) Impedance: 30 kΩ. (3) Reaction time: (2 ± 0.5) ms. (4) Resolution: 10 bits; accuracy: ± 0.6%; linearity: ± 0.15% of max. value
24 V	Logic input power supply	(min.: 15 V; max.: 30 V) Protected against short-circuits and overloads. Max. current available for customers 100 mA

(continued)

COM	Analog I/O common	0 V
L11 L12 L13 L14 L15	Programmable logic inputs	(1) +24 V_c (max. 30 V). (2) Impedance: 3.5 kΩ. (3) Reaction time: (2 ± 0.5) ms

Extension Materials

Common malfunction analysis

Scene I: Starter does not start, no fault displayed.

(1) If the display does not light up, check the connection to AL1, AL2 and RJ-45.

(2) The assignment of the "Fast stop" or "Freewheel stop" functions will prevent the drive starting if the corresponding logic inputs are not powered up. The ATV31 then displays [NST] in freewheel stop and [FST] in fast stop. This is normal since these functions are active at zero so that the drive will be stopped safely if there is a wire break.

(3) Make sure that the run command input or inputs are activated in accordance with the selected control mode.

(4) If an input is assigned to the limit switch function and this input is at zero, the drive can only be started up by sending a command for the opposite direction.

Scene II: Faults, which cannot be reset automatically (see Table 7.5).

Table 7.5 Faults, which cannot be reset automatically

Faults	Name	Probable cause	Remedy
CrF	[PRECHARGE FAULT]	Load relay control fault or charging resistor damaged	(1) Check the internal connections. (2) Inspect/repair the drive
EEF	[CONTROL EEPROM]	Internal memory fault	(1) Check the environment (electromagnetic compatibility). (2) Inspect/repair the drive
SCF	[MOTOR SHORT CCT]	(1) Short-circuit or grounding at the drive output. (2) Significant earth leakage current at the drive output if several motors are connected in parallel	(1) Check the cables connecting the drive to the motor, and the insulation of the motor. (2) Reduce the switching frequency. (3) Connect reactors in series with the motor

常见故障分析

情景Ⅰ：变频器不启动，无故障显示。

（1）如果显示器未点亮，检查与 AL1、AL2 和 RJ–45 的连接。

（2）如果没有启动相应的逻辑输入，则分配“快速停止”或“自由轮停止”功能将阻止驱动器启动。ATV31 然后显示 [NST] 在自由停止和 [FST] 在快速停止。这是正常的，因为这些函数处于零活动状态，因此如果有断线，驱动器将安全停止。

（3）确保根据选定的控制模式激活运行命令输入或输入。

（4）如果将输入分配给限位开关函数，并且该输入为零，则只能通过向相反方向发送命令来启动驱动器。

情景Ⅱ：不能自启动故障（见表 7.5）。

表 7.5　不能自启动故障

故障	名称	可能的原因	解决方案
CrF	[PRECHARGE FAULT]	负载继电器控制故障或者充电电阻损坏	（1）检查内部连接。 （2）检修 / 修理设备
EEF	[CONTROL EEPROM]	内部存储故障	（1）检查环境（电磁兼容性）。 （2）检修 / 修理设备
SCF	[MOTOR SHORT CCT]	（1）短路或设备输出接地。 （2）如果几个电动机并联，设备输出有明显的漏电流	（1）检查传动与电动机连接的电缆，电动机的绝缘情况。 （2）降低开关频率。 （3）将反应器与电动机串联

Self-Test

Translate the English (Table 7.6) **in the form into Chinese.**

Table 7.6　Power terminal and function

Power terminal（动力端子）	Function（功能）	Translation
⏚	Protective ground connection terminal	
R/L1 S/L2	Power supply	
R/L1 S/L2 T/L3		
PO	DC bus + polarity	
PA/+	Output to braking resistor (+ polarity)	
PB	Output to braking resistor	

(continued)

Power terminal（动力端子）	Function（功能）	Translation
PC/-	DC bus - polarity	
U/T1 V/T2 W/T3	Outputs to the motor	

New Words and Phrases

inverter *n.* 反相器；变频器
AC driver 变频器（欧美）
frequency converter 频率转换器；变频器（欧洲）
asynchronous motor 异步电动机
parameter *n.* 参数；参量；限制因素；决定因素
fault *n.* [电]故障
potentiometer *n.* 电位计
power indicated on plate 铭牌功率
apparent power 视在功率
power terminal 动力端子
braking resistor 制动电阻
impedance *n.* 阻抗，全电阻；电阻抗
solenoid valve 电磁阀，螺线管操纵阀
switching frequency 转换频率

Unit 8

Programmable Logic Controller

1. The brand of the mainstream PLC in the market (see Figure 8.1)

Figure 8.1 Various types of PLC

2. Describe PLC technique and future development

Programmable controller is the first in the late 1960s in the United States, then called PLC. Programmable logic controller is used to replace relays. For the implementation of the logic judgment, timing, sequence number, and other control functions. The concept is presented PLC general motor corporation. PLC and the basic design is the computer functional improvements, flexible, generic and other advantages and relay control system simple and easy to operate, such as the advantages of cheap prices combined controller hardware is standard and overall. According to the practical application of target software in order to control the content of the user procedures memory controller, the controller and connecting the accused convenient target.

In the mid-1970s, the PLC has been widely used as a central processing unit microprocessor, import/export module and external circuit are used, large-scale integrated circuits even when the PLC is no longer the logical (IC) judgment functions also have data processing PID conditioning and data communications functions. International Electro technical Commission (IEC) standards promulgated programmable controller for programmable controller draft made the following definition: programmable controller is a digital electronic computers operating system, specifically for applications in the industrial design environment. It used programmable memory, used to implement logic in their internal storage operations, sequence control, timing, counting and arithmetic operations, such as operating instructions, and through digital and analog input and output, the control of various types of machinery or production processes. Programmable controller and related peripherals, and industrial control systems easily linked to form a whole, to expand its functional design. Programmable controller for the user, is a non-contact equipment, the procedures can be changed to change production processes. The programmable controller has become a powerful tool for factory automation, widely popular replication. Programmable controller is user-oriented industries dedicated control computer, with many distinctive features.

3. Understanding how the S7-200 executes your control logic

(Taken from the manual of Siemens S7-200)

The S7-200 continuously cycles through the control logic in your program, reading and writing data.

1) The S7-200 relates your program to the physical inputs and outputs

The basic operation of the S7-200 is very simple:

(1) The S7-200 reads the status of the inputs.

(2) The program that is stored in the S7-200 uses these inputs to evaluate the control logic. As the program runs, the S7-200 updates the data.

(3) The S7-200 writes the data to the outputs.

Figure 8.2 shows a simple diagram of how an input electrical relay diagram relates to the S7-200. In this Start/Stop Switch example, the state of the switch for starting the motor is combined with the states of other inputs. The calculations of these states then determine the state for the output that goes to the actuator which starts the motor.

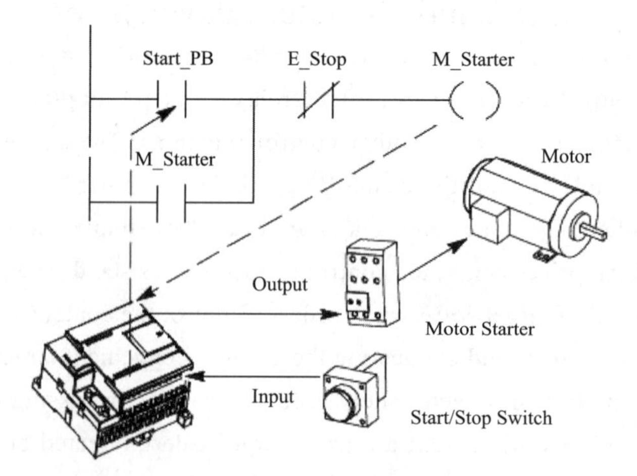

Figure 8.2 Controlling inputs and outputs

2) The S7-200 executes its tasks in a scan cycle

The S7-200 executes a series of tasks repetitively. This cyclical execution of tasks is called the scan cycle. As shown in Figure 8.2, the S7-200 performs most or all of the following tasks during a scan cycle:

(1) Reading the inputs: The S7-200 copies the state of the physical inputs to the process-image input register.

(2) Executing the control logic in the program: The S7-200 executes the instructions of the program and stores the values in the various memory areas.

(3) Processing any communications requests: The S7-200 performs any tasks required for communications.

(4) Executing the CPU self-test diagnostics: The S7-200 ensures that the firmware, the program memory, and any expansion modules are working properly.

(5) Writes to the outputs ; The values stored in process-image output register are written to the physical outputs.

The execution of the user program is dependent upon whether the S7-200 is in STOP mode or in RUN mode. In RUN mode, your program is executed; in STOP mode, your program is not executed.

3) Guidelines for designing a micro PLC system

There are many methods for designing a micro PLC system. The following general guidelines can apply to many design projects. Of course, you must follow the directives of your own company's procedures and the accepted practices of your own training and location.

(1) Partition your process or machine. Divide your process or machine into sections that have a level of independence from each other. These partitions determine the boundaries between controllers and influence the functional description specifications and the assignment of resources.

(2) Create the functional specifications. Write the descriptions of operation for each section of the process or machine. Include the following topics: I/O points, functional description of the operation, states that must be achieved before allowing action for each actuator such as solenoids, motors, and drives, description of the operator interface, and any interfaces with other sections of the process or machine.

(3) Design the safety circuits. Identify equipment requiring hard-wired logic for safety. Control devices can fail in an unsafe manner, producing unexpected startup or change in the operation of machinery. Where unexpected or incorrect operation of the machinery could result in physical injury to people or significant property damage, consideration should be given to the use of electro-mechanical overrides which operate independently of the S7-200 to prevent unsafe operations. The following tasks should be included in the design of safety circuits:

① Identify improper or unexpected operation of actuators that could be hazardous.

② Identify the conditions that would assure the operation is not hazardous, and determine how to detect these conditions independently of the S7-200.

③ Identify how the S7-200 CPU and I/O affect the process when power is applied and removed, and when errors are detected. This information should only be used for designing for the normal and expected abnormal operation, and should not be relied on for safety purposes.

④ Design manual or electro-mechanical safety overrides that block the hazardous operation independent of the S7-200.

⑤ Provide appropriate status information from the independent circuits to the S7-200 so that the program and any operator interfaces have necessary information.

⑥ Identify any other safety-related requirements for safe operation of the process.

4) Specify the operator stations

Based on the requirements of the functional specifications, create drawings of the operator stations. Include the following items:

(1) Overview showing the location of each operator station in relation to the process or machine.

(2) Mechanical layout of the devices, such as display, switches, and lights, for the operator station.

(3) Electrical drawings with the associated I/O of the S7-200 CPU or expansion module.

5) Create the configuration drawings

Based on the requirements of the functional specification, create configuration drawings of the control equipment. Include the following items:

(1) Overview showing the location of each S7-200 in relation to the process or machine.

(2) Mechanical layout of the S7-200 and expansion I/O modules (including cabinets and other equipment).

(3) Electrical drawings for each S7-200 and expansion I/O module (including the device model numbers, communications addresses, and I/O addresses).

6) Create a list of symbolic names (optional)

If you choose to use symbolic names for addressing, create a list of symbolic names for the absolute addresses. Include not only the physical I/O signals, but also the other elements to be used in your program.

(From *S7-200 Programmable Controller System Manual*)

Extension Materials

The Development of PLC controllers

Industry has begun to recognize the need for quality improvement and increase in productivity in the sixties and seventies. Flexibility also became a major concern (ability to change a process quickly became very important in order to satisfy consumer needs). Try to imagine automated industrial production line in the sixties and seventies. There was always a huge electrical board for system controls, and not infrequently it covered an entire wall. Within this board there were a great number of interconnected electromechanical relays to make the whole system work. By word "connected", it was understood that electrician had to connect all relays manually using wires. An engineer would design logic for a system, and electricians would receive a schema often contained hundreds of relays. The plan that was given was by electrican called "ladder schematic". Ladder displayed all switches, sensors, motors, valves, relays, etc. found in the system. Electrician's job was based on mechanical relays. Mechanical instruments were usually the weakest connection in the system due to their moveable parts that could wear out. If one relay stopped working, electrician would have to examine an entire system (system would be out until a cause of the problem was found and corrected).

The other problem with this type of control was in the system's break period when a system had to be turned off, so connections could be made on the electrical board. If a firm decided to change the order of operations (make even a small change), it would turn out to be a major expense and a loss of production time until a system was functional again.

It's not hard to imagine an engineer who makes a few small errors during his project. It is also conceivable that electrician has made a few bad components. The only way to see if everything is all right is to run the system. As systems are usually not perfect with a first try, finding errors was an arduous process. You should also keep in mind that a product could not be made during these corrections and changes in connections. System had to be literally disabled before changes were to be performed. That meant that the entire production staff in that line of production was out of work until the system was fixed up again. Only when electrician was done finding errors and repairing, the system was ready for production. Expenditures for this kind of work were too great even for well-todo companies.

"General Motors" is among the first who recognized a need to replace the system's "wired" control board. Increased competition forced auto-makers to improve production quality and productivity. Flexibility and easy change of automated lines of production became crucial! GM's idea was to use for system logic one of the microcomputers (these microcomputers were as far as their strength beneath today's eight-bit microcontrollers) instead of wired relays. Computer could take place of huge, expensive, inflexible wired control boards. If changes were needed in system logic or in order of operations, program in a microcomputer could be changed instead of rewiring of relays. Imagine only what elimination of the entire period needed for changes in wiring meant then. Today, such thinking is but common, and then it was revolutionary!

Everything was well thought out, but then a new problem came up of how to make electricians accept and use a new device. Systems are often quite complex and require complex programming. It was out of question to ask electricians to learn and use computer language in addition to other job duties. General Motors company recognized a need and wrote out project criteria for first programmable logic controller (there were companies which sold instruments that performed industrial control, but those were simple sequential controllers and not PLC controllers as we know them today). Specifications required that a new device be based on electronic instead of mechanical parts, to have flexibility of a computer, to function in industrial environment (vibrations, heat, dust, etc.) and have a capability of being reprogrammed and used for other tasks. The last criterion was also the most important, and a new device had to be programmed easily and maintained by electricians and technicians. When the specification was done, General Motors looked for interested companies and encouraged them to develop a device that would meet the specification for this project.

"Gould Modicon" developed a first device which met these specifications. The key to success with a new device was that for its programming you didn't have to learn a new programming language. It was programmed so that same language this ladder diagram, already know to technicians because the logic looked similar to old logic that they were used to working with. Thus they didn't have to learn a new programming language which (obviously) proved to be a good move. PLC controllers were initially called PC controllers (programmable controllers). This caused a small confusion when personal computers appeared. To avoid confusion, a designation PC was left to computers, and programmable controller became programmable logic controllers. First PLC controllers were simple devices. They connected inputs such as switches, digital sensors, etc. , and based on internal logic they turned output devices on or off. When they first came up, They were not quite suitable for complicated controls such as temperature, position, pressure, etc. However, throughout years, makers of PLC controllers added numerous features and improvements. Today's PLC controller can handle highly complex task such as position control, various regulation and other complex applications. The speed of work and easiness of programming were also improved. Also, modules for special purpose were developed, like communication modules for connecting several PLC controllers to the net. Today it is difficult to imagine a task that could not be handled by a PLC.

PLC 控制器的发展

在 20 世纪 60 年代和 70 年代，工业领域已经开始意识到提高产品质量和生产力的必要性。灵活性也是一个主要问题（为了满足消费者的需求，快速改变过程的能力变得非常重要）。试想一下 20 世纪 60 年代和 70 年代的自动化工业生产线。往往使用一个巨大的电气板用于系统控制，而且覆盖整个墙壁。在这个板内有大量相互连接用于整个系统工作的继电器。根据“连接”一词所理解，电工必须使用导线手动连接所有继电器。一个工程师设计一个系统逻辑，电工需要数百个继电器来完成。电工给出的文案称为“梯形图”。图中显示系统中所有开关、传感器、电动机、阀门、继电器等。电工的工作是针对机械继电器从机械设备的可移动部件会磨损,所以是系统中连接最不稳定的。如果一个继电器停止工作,电工将不得不检查整个系统（系统在发现和纠正问题之前都是停止的）。

这种类型控制的另一个问题是当系统必须关闭时所面临的系统中断期，因为所有连接是在一块电路板上进行的。如果一家公司决定改变操作顺序（即使是一个小的变化），在系统再次运行前，会产生大量的费用和生产时间的损失。

不难想象一个工程师在他的项目中会犯一些小错误。也可以想象到电工也会使用一些坏的器件。运行系统是判断一切是否正常的唯一方法。由于系统首次运行通常是不完善的，发现错误是一个艰难的过程。你应该还记住，在维修和更改连接期间是不能进行产品生产的。在更改完成之前,系统必须完全禁用。这意味着系统再次修复前生产线上的所有生产人员都失业了。只有当电工发现错误并排除故障时，系统才能进行生产。即使对资金雄厚的公司来说，这类工作的开支也太大了。

“通用汽车”是第一批认识到需要更换系统的“有线”控制板的人之一。竞争加剧迫使汽车制造商提高生产质量和生产率。自动化生产线的灵活性和易变性变得至关重要！ 通用汽车的想法是将其中一台微型计算机用于系统逻辑（这些微型计算机的强度远远低于现在的 8 位微控制器），代替有线的继电器。计算机可以取代巨大的、昂贵的、不灵活的有线控制板。如果系统逻辑或操作顺序需要更改,则可以更改微机中的程序,而不是重新布线继电器。想象一下，为了改变线路，所需消耗的整个时间意味着什么。现在，这样的想法是常见的，然而在当时是革命性的!

一切都考虑得很好,但后来出现了一个新的问题,即如何使电工接受和使用一个新的设备。系统往往相当复杂，且需要复杂的编程。除了其他工作职责外，要求电工学习和使用计算机语言是毫无疑问的。通用汽车公司意识到了需要一个可编程逻辑控制器，并写出了项目标准（有些公司出售进行工业控制的仪器，但这些是简单的顺序控制器，而不是我们所知道的 PLC 控制器）。规范要求一种新的设备以电子而不是机械部件为基础，它具有计算机的灵活性，能在工业环境（振动、热、灰尘等）中发挥作用，并具有被重新编程和用于其他任务的能力。最后一个标准也是最重要的，新设备必须易于编程，并由电工和技术人员维护。当规范完成后，通用汽车公司寻找感兴趣的公司，并鼓励他们开发一种符合本项目规范的设备。

“古尔德莫迪康”开发了第一个符合这些规范的设备。一个新设备成功的关键是你不必学习一种新的编程语言。它使用梯形图进行编程，技术人员已经知道，因为逻辑看起来类似于他们曾经使用的旧逻辑。因此，他们不必学习一种新的编程语言，这（显然）被证明是一个很好的举动。PLC 控制器最初称为 PC 控制器（可编程控制器）。当个人计算机出现时，这引起了小小的混乱。为了避免混淆，指定 PC 留给计算机使用，可编程控制器称为可编程逻辑控制器。首先，PLC 控制器是简单的设备。它们连接输入，如开关、数字传感器等。根据内部逻辑，它们打开或关闭输出设备。当它们刚开始出现时，不太适合于复杂的控制，如温度、位置、压力等。然而，多年来，PLC 控制器的制造商增加了许多功能并进行改进。现在的 PLC 控制器可以处理高度复杂的任务，如位置控制、各种调节和其他复杂的应用。也提高了工作速度和编程的简易性。此外，还开发了特殊用途的模块，如用于将几个 PLC 控制器连接到网络的通信模块。现在，很难想象会遇到一个 PLC 无法处理的任务。

Self-Test

You should depend on the dictionary to finish the reading, and fill on the blank with your own thoughts (Figure 8.3).

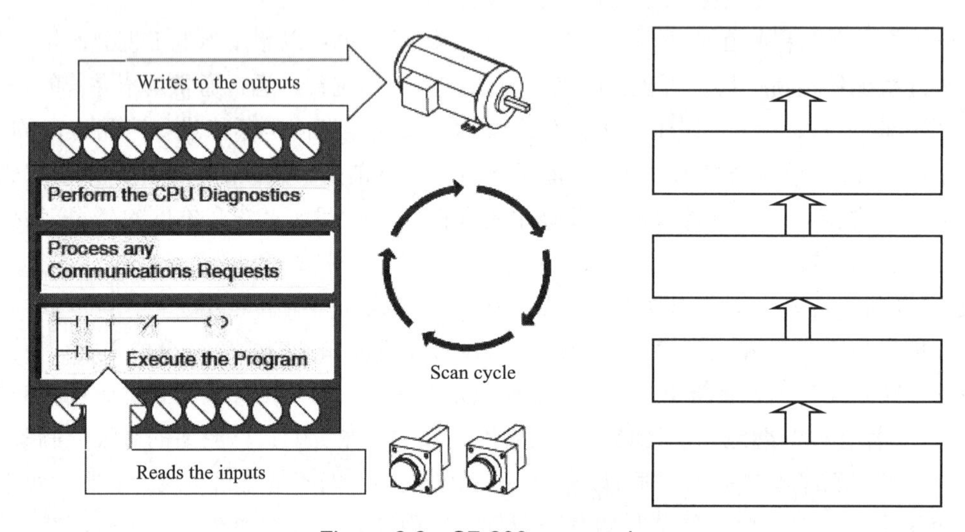

Figure 8.3 S7-200 scan cycle

New Words and Phrases

continuously *adv.* 连续不断地，接连地；时时刻刻

relate to 涉及；同……有……关系；与……协调；将……与……联系起来

be combined with 与……结合

actuator *n.* 执行机构；激励者；执行器

repetitive *adj.* 重复的，反复的

cyclical *adj.* 循环的；周期的；环状的

process-image input register 输入映像寄存器

diagnostics *n.* 诊断

firmware *n.* 固件（软件硬件相结合）；稳件

converter *n.* [自]变换器；换流器，变频器；[冶]炼钢炉，吹风转炉

functional description 功能描述

solenoid *n.* 螺线管；线包

electro-mechanical *adj.* 电动机械的，机电的，电机的

layout *n.* 布局，安排，设计；布置图，规划图

drawing *n.* 制图

cabinet *n.* 控制柜

optional *adj.* 可选择的；随意的，任意的；非强制的

stimulus *n.* 激励

pushbutton *n.* 按钮

thumbwheel *n.* 指轮

hardwired *adj.* 硬连接的

rewiring *n.* 重新布线，重新接线

repairable *adj.* 可修复的

module *n.* 模块

variable-speed drive 变速传动装置

traffic director 交通控制器

scan cycle 扫描周期

instructions of the program 程序指令

self-test diagnostics 自诊断

Unit 9

Pneumatic Control System

1. The pneumatic control devices (see Figure 9.1)

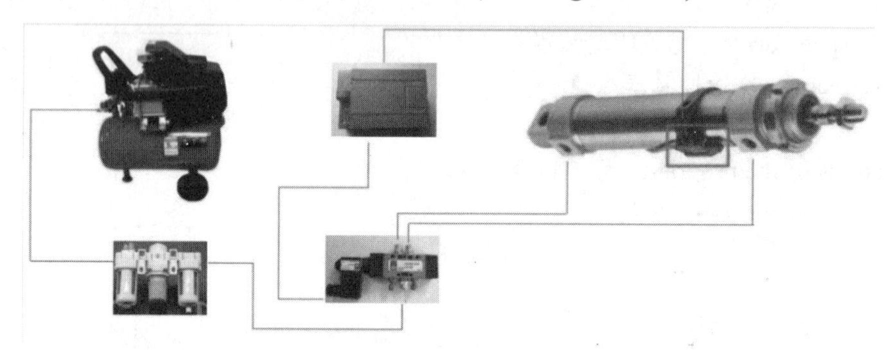

Figure 9.1 Schematic diagram of pneumatic control devices

Pneumatic power can be supported with the aid of compressors such as the centrifugal, the axial-flow and the positive-displacement types.

1) Air compressor (see Figure 9.2)

Figure 9.2 Air compressor and the nameplate

2) Pneumatic triplet (see Figure 9.3)

Figure 9.3 Pneumatic triplet and nameplates

3) Valves

Valves are used to control the direction and amount of flow in a pneumatic system, as shown in Figure 9.4. A number of different valves may need to be used on pneumatic conveying plant, and a wide variety of different valves are available in the market place. Rotary valves have been considered at length, and are ideal for controlling the feed of material into or out of a system at a controlled rate. There is, however, a requirement for many other types of valve, generally to be used for the purpose of isolating the flow. Such as discharge valves, vent line valves and diverter valves.

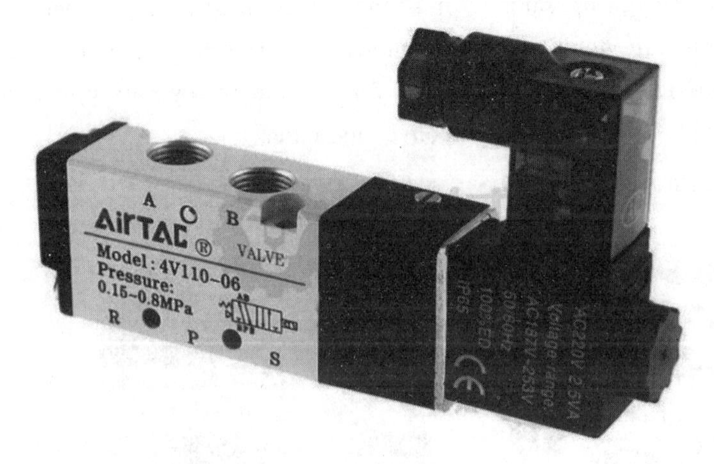

Figure 9.4 Electromagnetic valve

(1) Discharge valves. A valve in a conveying line that is required to stop and start the flow is an onerous duty. Although the valve is only used in either the open or closed position, and is not used for flow control purposes, particulate material must be able to pass freely through when it is open. If the control surfaces of the valve remain in the flow path, as they will with pinch valves and ball valves, they must provide a perfectly smooth passage for the flow of material through the valve when open.

Any small protuberances or surface irregularities that could promote turbulence in the area would result in a rapid deterioration in performance. This is particularly the case when the material to be conveyed is abrasive. This type of valve is also very vulnerable during the opening and closing sequences, and so these operations should be completed as quickly as possible.

(2) Vent line valves. This is a deceptively easy duty, but if it is on a high pressure blow tank handling a material such as fly ash or cement, the valve will have to operate in a very harsh environment. With the venting of high pressure air the air velocity will be very high, albeit for a very short period of time. As a consequence of the turbulence in the blow tank, however, a considerable amount of abrasive dust is likely to be carried with the air. If the material is abrasive then the choice is between a pinch valve and a dome valve. If the material is non-abrasive, a diaphragm valve could be used.

(3) Diverter valves. There are two main types of diverter valves. In one, a hinged flap is

located at the discharge point of the two outlet pipes. This flap provides a seal against the inlet to either pipe. The pipe walls in the area are lined with urethane, or similar material, to give an airtight seal, and this provides a very compact and light weight unit. The author tested a Y-branched diverter valve of this design with silica sand in dilute phase, but it was a disaster. After conveying only 12 tonne of sand, the 4 mm thick bronze flap had a 15 mm diameter hole through it. The urethane lining, however, was in perfect condition. The problem was that the sand was always impacting against the flap. A straight through design with a branch off would have been better, but still not suitable for abrasive materials.

The other main design operates with a tunnel section of pipe between the supply and the two outlet lines. This unit would not be recommended for abrasive materials either. This design, however, should provide a more positive seal for the line not operating, which would probably make it a more suitable valve for vacuum conveying duties.

4) Air cylinders (see Figure 9.5)

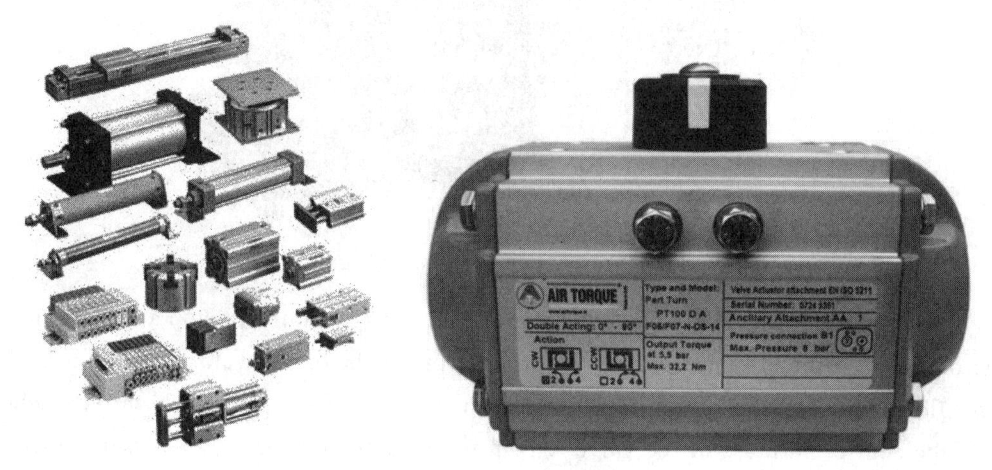

Figure 9.5 Air cylinders

2. The pneumatic control system

Figure 9.6 gives the layout of the pneumatic circuit. The explanation of what happens during the cushioning is easy if the results of a digital simulation are available that show the piston velocity. The simulation model is shown in Figure 9.7. One of the main reasons for digital simulation of pneumatic systems the ease with which parameters can be changed either manually or programmed. In this case, the setting of the cushion screw was changed from fully open to partially open, from a sonic conductance of 100% to 40%.

The symbol in Figure 9.8 shows barrel, piston, and rod. The arrow and the two rectangles beside the piston symbolize the adjustable cushioning. Cylinders convert pneumatic energy to mechanical work. They usually consist of a movable element such as piston and piston rod, or plunger, operating within a cylindrical bore. Cylinders are often double-sided, i.e. pressurised air can work on both side of the piston to extend or retract it, and they have mostly a single-ended piston rod.

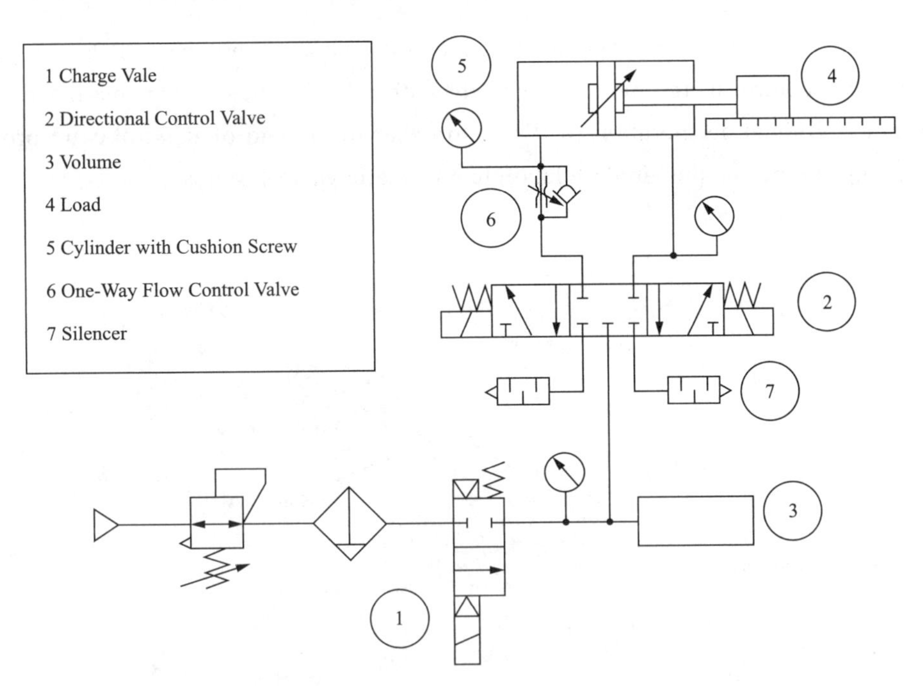

Figure 9.6 Pneumatic circuit of test rig

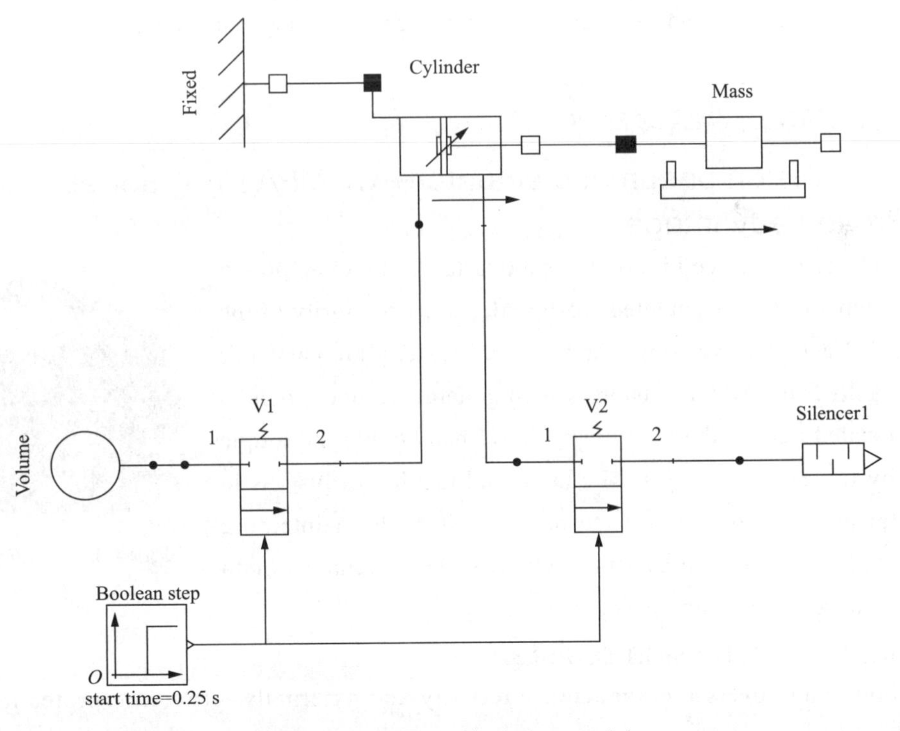

Figure 9.7 The simulation model

A typical design is shown in Figure 9.8. The piston rod is case hardened and chrome plated while the barrel is made of stainless steel or—for tie-rod cylinders—of an aluminium profile. Most cylinders have a band of magnetic material around the circumference of the piston and are fitted with a non-magnetic cylinder barrel. The magnetic field will travel with the piston as the piston rod moves in and out. By placing magnetically operated switches on the outside of the barrel, electronic control of the piston movement with a PLC is possible. Some means of stroke cushioning, i.e. gradual deceleration of the piston near to the end of its stroke are provided by cushioning rings in the end position or elaborate pneumatic valve systems.

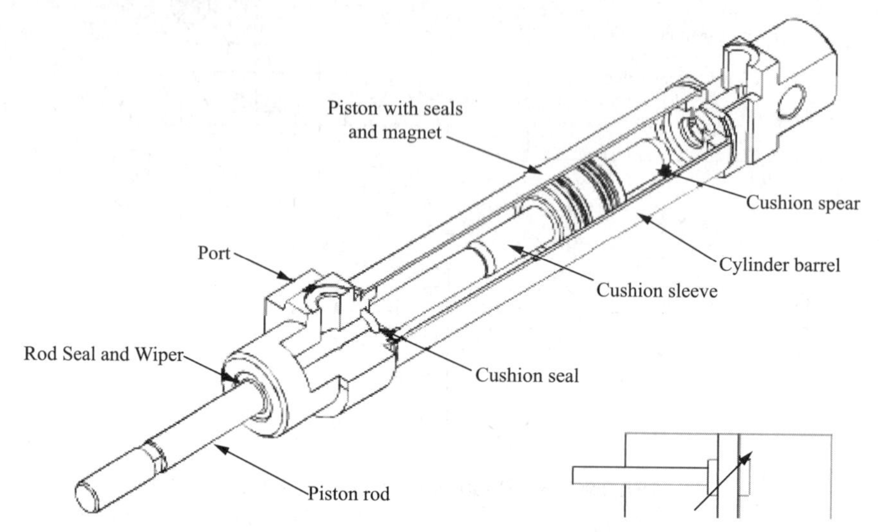

Figure 9.8 Cut-away view of single rod cylinder according to ISO 6432 and symbol

Extension Materials

Piston pneumatic actuator—COMPACT Ⅱ series

1. Proven advantage

The COMPACT Ⅱ see Figure 9.9 is a quarter turn rack & pinion pneumatic actuator that is patented worldwide. The superiority of the COMPACT Ⅱ actuator over single and double rack & pinion actuator designs, results from the four pistons which generate torque around a centrally located pinion, thereby giving more than double the torque achieved by these other designs. The increased number of pistons in the actuator allows their diameter to be reduced while maintaining its high torque. This also allows the overall size of the actuator to reduces and become more compact.

Figure 9.9 COMPACT Ⅱ

2. Superior corrosion resistance

The body and covers are anodized internally and externally, providing protection against ingress of corrosive atmosphere.

An external epoxy base layer and a second polyurethane paint provides additional protection against aggressive environments. Optional Electroless Nickel Coating of body, covers and stop.

3. Balanced forces

The cube-shaped configuration of the COMPACT Ⅱ positions the pistons so that each piston develops thrust along its own axis, rather than the off-axis thrust, that results from the geometry of most other actuator configurations. Piston side loading, caused by off-axis thrust, does not occur, thus resulting in less stress on the seals.

4. Space saving, fast acting

The COMPACT Ⅱ has four small cylinders each located on one of the four sides of a cube. At a given air pressure, the COMPACT Ⅱ can produce the same torque output as double piston actuators, using smaller diameter pistons and a narrower pinion, A narrower pinion results in a shorter piston travel, which permits a compact, space saving mechanism and fast acting travel from one position to the next.

5. Less air consumption

The COMPACT Ⅱ gives maximum torque for minimum air consumption. It is both compact in size and energy efficient, creating a fast-responding, trouble-free, high cycle lifespan.

The cube shape of the COMPACT Ⅱ and the short piston travel serve to minimize excess space. This is space which is not swept by piston travel and which must be pressurized before the piston motion begins; therefore, reducing the pressurization of excess space and resulting in reduced energy requirements.

6. Nested springs

The COMPACT Ⅱ four-spring chambers can use up to three different spring sizes, which are nested between the covers and pistons and are aligned by centering rings. Each spring is wound in the opposite direction to its neighbor to avoid entanglement. As there are four cylinders, there are many more spring combination possibilities than with double piston actuators. This results in better solutions for any air supply pressure required. Special painting of the springs provides higher corrosion resistance to the environment, giving more than 250 hours of life in a salt spray bath.

7. Less wear

With its unique 4-piston design, the COMPACT Ⅱ achieves a more uniform load distribution than do single and double piston actuators, therefore greatly reducing gear wear at the points of contact between rack and pinion. The force-balanced piston with its shorter stroke prevents uneven wear of O-rings, gear and pistons. This eliminates the number of soft parts, thereby resulting in longer maintenance schedules and low cost of repair kits. The high surface fish of the four-cylinder is protected from wear due to the hardened surface created by the anodizing treatment.

8. Limit stop (see Figure 9.10)

The pinion and stop rotation can be adjusted by four large diameter adjustable set screws diametrically opposed and threaded into the actuator body. Each opposing pair of screws exerts simultaneous and equal forces on opposite side of the stop when the rotation limit is reached, thus, no off-center forces are generated.

The stop allows for $\pm 5^\circ$ of rotational adjustment in both directions of travel. Any intermediate position can be achieved with a longer set of stop screws. This feature is built into the actuator stop mechanism and eliminates the need for additional plates and screws.

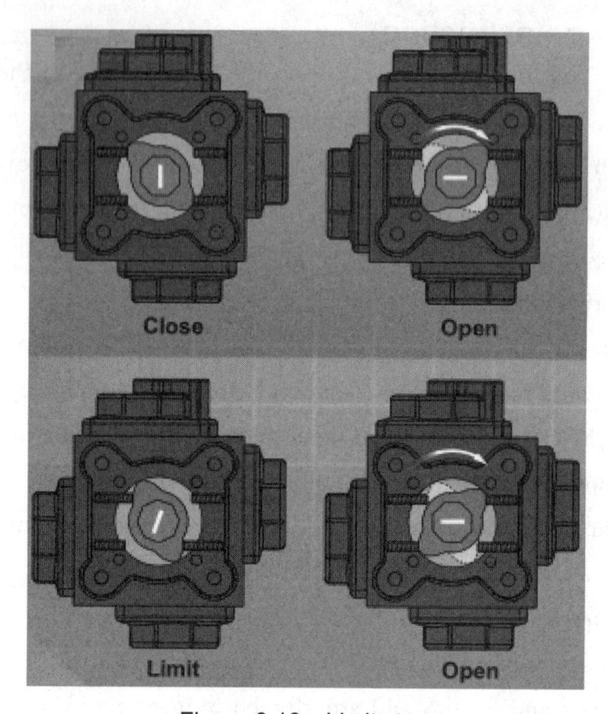

Figure 9.10 Limit stop

9. Indicator & Puck

A highly visible indicator with flow direction arrows is snapped to the pinion providing easy identification of valve position. The indicator snap-on arrows allow true position of any type of valve porting.

A puck with three position signaling inserts and a highly visible indicator with flow direction arrows is bolted to the pinion to provide a cost effective option for valve monitoring.

10. Pinion

The pinion has a double-square female drive on its bottom plane for accepting the ISO 5211 or DIN 3337 coupling options. The top plane has the Namur slot for attachment to switches or positioners. There is a machined flat below the NAMUR interface to provide for manual operation of the actuator by use of a wrench. The pinion is made from carbon steel with EN plating which gives a hard wearing surface with added protection against corrosive environments.

The COMPACT Ⅱ actuator transforms the linear motion of it's pneumatic pistons into rotary motion via 4 gear racks that drive the central pinion. Air Supply, to drive the pistons, flow into port A of the NAMUR cover: port A is connected to the center chamber and port B is connected to the four outside chambers.

11. Double acting (see Figure 9.11)

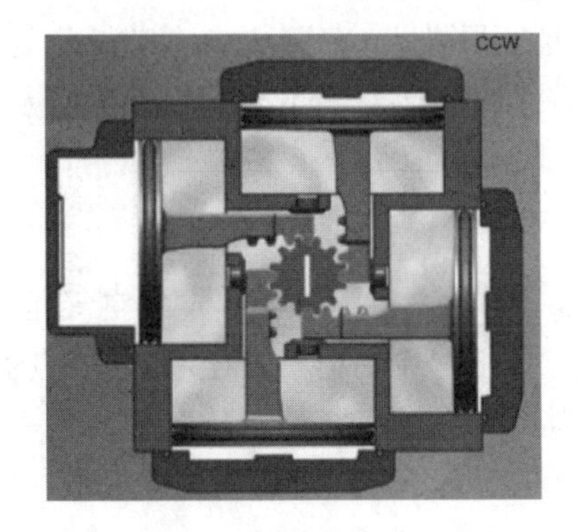

Center chamber pressurized; Pistons move outward; Pinion rotates counter clockwise

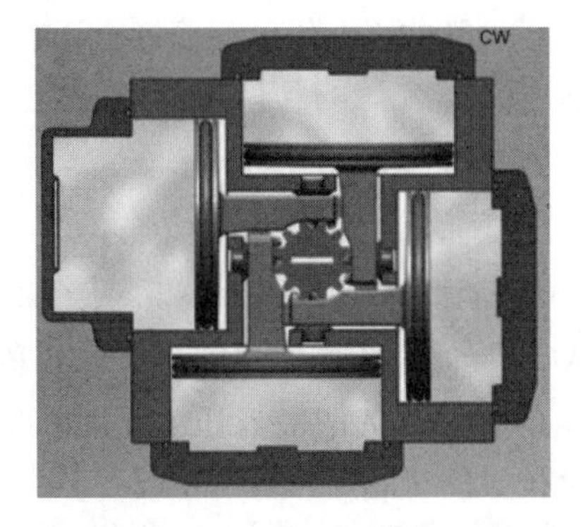

Outside chambers pressurized; Pistons move inward; Pinion rotates clockwise

Figure 9.11 Double acting

12. Spring return (see Figure 9.12)

Center chamber pressurized. pistons move outward. Springs are compressed. Pinion rotates counter clockwise

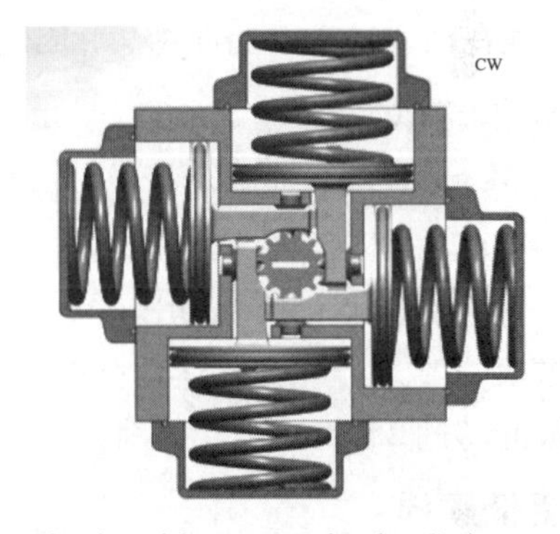

Air released from center chamber. Springs drive pistons inward. Pinion rotates clockwise

Figure 9.12 Spring Return

活塞气动执行器——COMPACT Ⅱ系列

1. 已证实的优点

该COMPACT Ⅱ（见图9.9）是已获专利的四分之一转架和小齿轮气动执行器。该COMPACT Ⅱ执行器优于单齿轮和双齿轮驱动器设计，这是由于围着中心位小齿轮的四个活塞产生扭矩，从而给予超过其他设计两倍的扭矩。执行器中增加的活塞数量允许直径减小的同时保持其高扭矩。这也使执行器的整体尺寸减小并变得更加紧凑。

2. 耐腐蚀性能优越

阀体和阀盖内部和外部进行阳极氧化，以防止腐蚀性大气的进入。

外部环氧基层和第二聚氨酯涂料提供额外的保护，以防止侵蚀性环境。可选化学镍涂层覆盖以阻止腐蚀。

3. 平衡力

COMPACT Ⅱ的立方体形状使每个活塞沿着自己的轴发展推力，而不是轴外推力，这是由大多数其他执行器配置的几何形状形成的。由离轴推力引起的活塞侧不会产生负载，从而导致密封的应力较小。

4. 节省空间，速度快

该 COMPACT Ⅱ有四个小圆柱体，每个位于一个四方的立方体中。在给定的空气压力下，该 COMPACT Ⅱ能产生等同于双活塞执行器的相同扭矩输出，其使用小直径活塞和窄的小齿轮，齿轮窄导致更短的活塞行程，而这促成一个更紧凑、节约空间机制以及运行时从一个位置到另一个位置的快速反应。

5. 更少的空气消耗量

COMPACT Ⅱ可给出最大扭矩而需求最低空气消耗量。它体积小巧、节能、反应迅速、无故障、长循环寿命。

立方体形状的 COMPACT Ⅱ和短活塞行程有利于减少多余的空间。不被活塞移动扫过的空间，必须在活塞运动开始前增压，因此，减少多余空间的增压，可达到减少能源需求的效果。

6. 嵌套的弹簧

COMPACT Ⅱ的“四弹簧室”可以使用三个不同大小的弹簧，表面和活塞之间的嵌套和定心环是一致的。每个弹簧都绕着与其相邻弹簧相反的方向旋转，以避免碰撞纠缠。由于有四个气缸，导致其比双活塞驱动器有更多的弹簧组合。这亦将为任何需要的空气供应压力提供更好的解决方案。特殊弹簧漆为弹簧提供了更高的耐腐蚀性环境，在专业盐雾浴喷洒实验环境下可保证超过 250 h 的使用寿命。

7. 更低磨损

以其独特的四活塞设计，相较于单活塞或双活塞执行器，COMPACT Ⅱ可达成更均匀载荷分布，因此这大大降低齿轮齿条与齿轮之间的接触点磨损。具有更短冲程的力平衡活塞防止 O 形环、齿轮和活塞的不均匀磨损。这消除了软部件的数量，从而导致更长的维护时间和维修套件的低成本。四活塞设计的气缸表面被保护，这样免于因为阳极处理后表面硬化带来的磨损。

8. 限制器（见图 9.10）

小齿轮和停止旋转可以通过四个直径可调的固定螺钉直径相对并插入执行器内进行调整。当达到旋转极限时，每一对相对的螺钉同时对停止装置的另一侧施加相等的力，因此不会产生偏心力。

该限制器允许 $\pm 5^{\circ}$ 范围内的旋转调整并在两个方向的运行。任何中间位置都可以用一组更长的制动螺钉来实现。这种功能内置在执行器的停止机制中，不需要额外的钢板和螺钉。

9. 指示器和定标器

一个带有流向箭头的高度可见的指示器被拉到小齿轮上，以便于识别阀门的位置。指示器上的可脱卸箭头允许任何类型的阀门端口的真实位置。

齿轮上栓有一个带三个位置信号插入件的定标器和一个高度可见的带有流向箭头的指示器，为阀门监控提供了一种经济有效的可行选择。

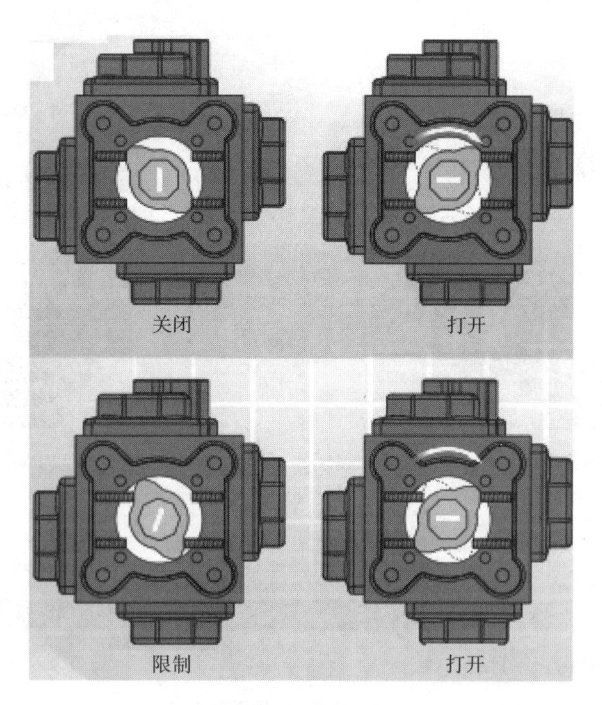

图 9.10　限制器

10. 小齿轮

小齿轮在其底部平面上有一个双方形母驱动器，用于接受 ISO 5211 或 DIN 3337 联轴器。顶部平面有 NAMUR 槽，用于连接到开关或定位器。Namur 接口下方有一个机加工平面，供使用扳手手动操作执行机构。小齿轮是由碳钢用电镀制成，提供了一个坚硬的磨损表面和附加的保护，以防止腐蚀环境。

COMPACT Ⅱ执行器通过 4 架中央小齿轮驱动将气动活塞的直线运动转化为旋转运动。驱动活塞的供气进入 NAMUR 盖的 A 口：A 口与中心腔室相连，而 B 口与四个外部腔室相连。

11. 双动式（见图 9.11）

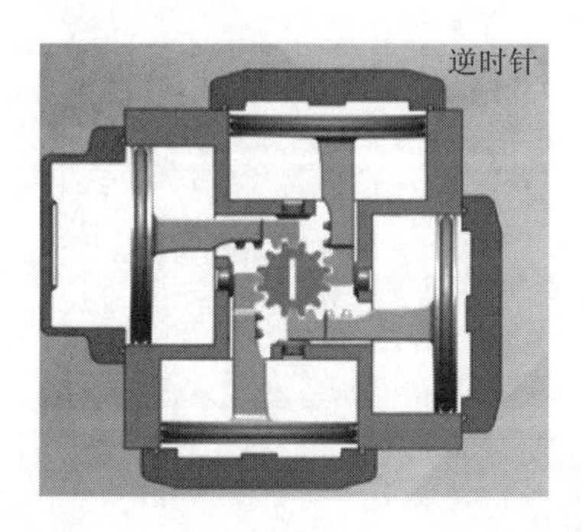

中心室加压；活塞移动向外；小齿轮逆时针旋转

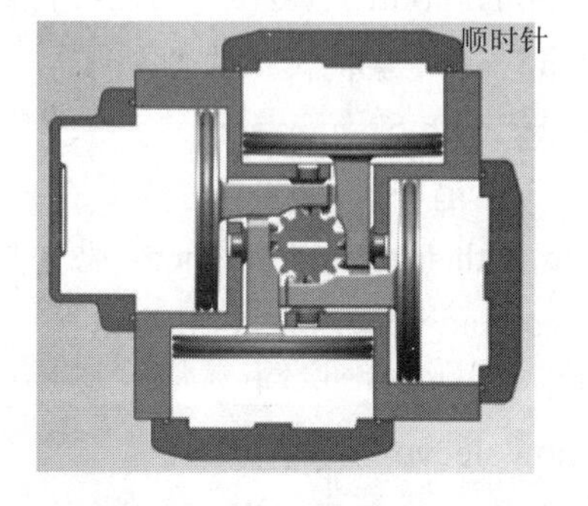

外室加压；活塞移向内；小齿轮旋转顺时针

图 9.11　双动式

12. 弹簧还原（见图 9.12）

中心室加压；活塞移动向外；弹簧被压缩；小齿轮逆时针旋转

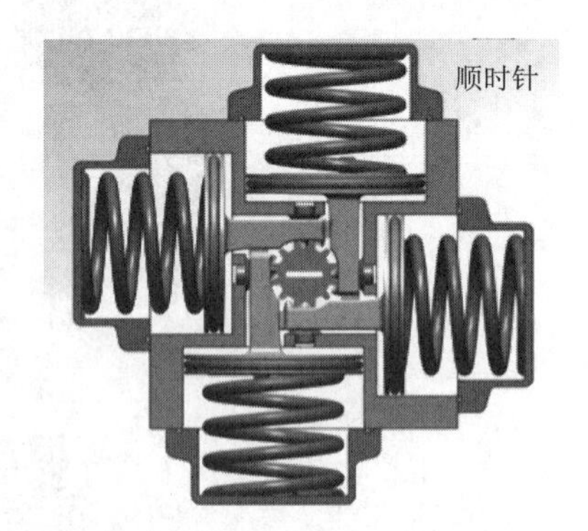

空气从中心室排出；弹簧向内驱动活塞；小齿轮顺时针旋转

图 9.12 弹簧还原

Self-Test

Translate the words into English.

(1) 气缸。

(2) 空气压缩机。

(3) 电磁阀。

(4) 油雾器。

(5) 减压阀。

(6) 空气过滤器。

New Words and Phrases

pneumatic *adj.* 充气的；气动的

charge valve 充气阀，加载阀，加液阀

directional control valve 方向控制阀，定向控制阀

volume *n.* 量，大量；体积

cylinder with cushion screw 带转角螺钉的圆筒

silencer 消音器

piston with seals and magnet 带密封件和磁铁的活塞

rod seal and wiper 棒封和刮水器

cushion seal 缓冲气垫

cushion sleeve 缓冲套

cylinder barrel 气缸筒

cylinder spear 气缸杆

CHAPTER 4

Power System Protection

Objectives:

(1) Use your own background information to define a power supply and protection.

(2) Know the various protective devices.

Unit 10

Power System Protection

1. The potential dangers of the power system

The protection requirement in an electric circuit depends on the anticipated hazards relative degree of protection required against each and the capacity of the undertaking to bear the protective devices. Protection is provided to minimize the following hazards in the system:

(1) Sustained excess current, i.e., overloading.

(2) Excess energy during transient overload, i.e., short circuit faults, etc.

(3) Electric shock.

(4) Fire.

(5) Loss of discrimination.

(6) Interruption of supplies.

At the same time, protection is required to be effective with the least disturbance to the normal operation of the system, both in the branch protected as well as in the remaining branches. In other words, there should be no avoidable loss of discrimination at any level of fault current, the nuisance created by the interruption of supplies should be kept to a minimum and the restoration of supply on the clearance of a fault should be as rapid as possible. On immediate inspection, it is clear that some of these requirements are to a large extent contradictory and a compromise must be reached in a practical situation.

2. The protective devices for the power system

1) Fuse

(1) Introduction: A fuse is a combination of a protective device and a circuit breaker (see Figure 10.1). The time/current characteristic of a fuse is approximately $I^2 \cdot t$=constant for large currents, so that its operating time can be very small (fraction of a cycle) and thus the time grading is not a simple matter. Current-time characteristics of different fuse links and elements differ and need careful consideration for application in a system. Fuses can be of low or high rupturing capacity. Rewirable (可更换熔丝的) type are considered having low breaking capacity. All fuses having a breaking capacity of over 16 kA can be referred to as high rupturing capacity (HRC) (高切断能力) fuses as specified in IS:2208—1962, HRC fuses provide better discriminative protection to a degree of sensitivity which is well within the requirements of general distribution. Their characteristics can be ascertained and declared accurately so that they can be applied in situations where the requirements are known. These remain consistent and stable in service without calibration or maintenance, provided they are properly designed and manufactured. The design of

the system in which fuses are used and other factors external to the fuse itself, affect the degree of discrimination which can be achieved by possible matching of current-time characteristics of the fuse link as well as of the device to be protected. An assessment of these factors is necessary for the most effective use of fuses, though the assessment need not be critical. Failure to achieve discrimination owing to the incorrect choice of fuses may result in a nuisance.

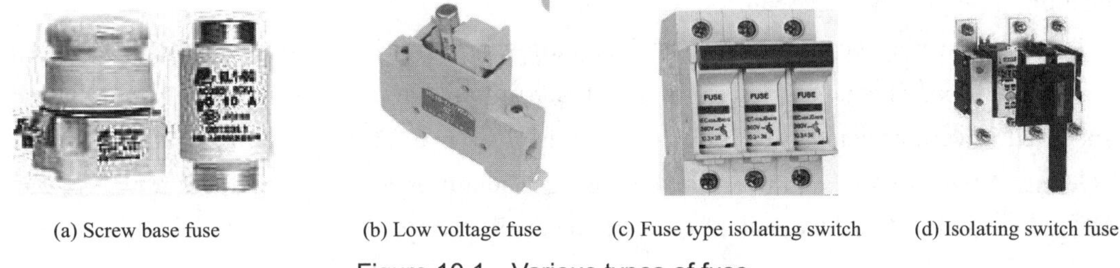

(a) Screw base fuse (b) Low voltage fuse (c) Fuse type isolating switch (d) Isolating switch fuse

Figure 10.1 Various types of fuse

(2) To know the high voltage fuses. Figure 10.2 shows various types and application of high-voltage fuse.

Figure 10.2 Various types and application of high-voltage fuse

There are of two types as described below.

① current-limiting fuses.These fuses which are of high rupturing capacity are silica powerfilled, contained in a porcelain envelope with an entire assembly of elements on the former and is impressed in the file which may be air or oil tight depending upon the application desired (see Figure 10.3).

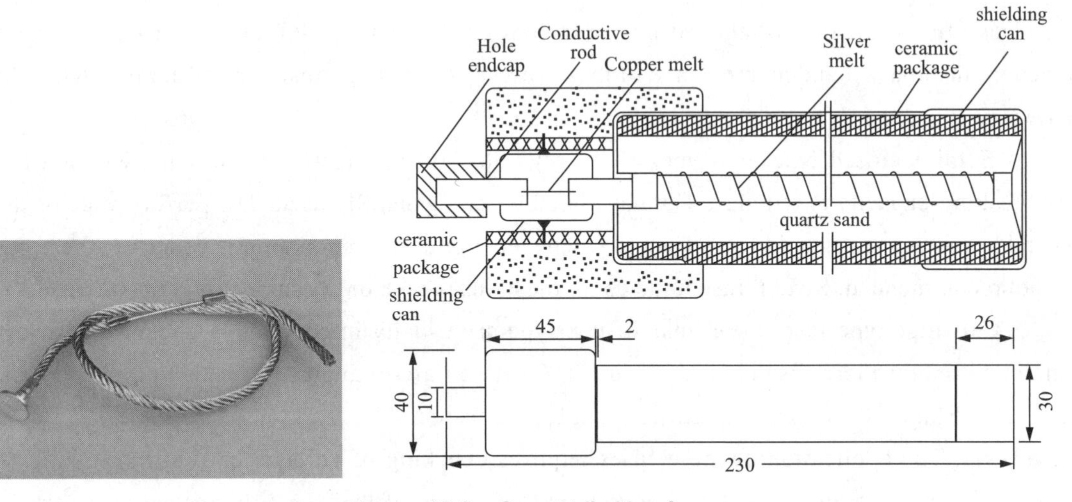

Figure 10.3 Current-limiting fuse

Rewirable types fuses: horn gap type. These are of a lower rupturing capacity.

The horns are generally made of steel. For effective quenching of the arc the angle of dispersion of the arcing horns is designed for 60° and fuse assembly is installed in a horizontal position. Tinned copper wire is used generally as the fuse element. The working of these fuses is effected due to pollution, wind and other atmospheric conditions and hence they do not afford complete protection, though they are cheap.

The element may be enclosed in a glass tube when it is liable to mechanical damage in an open air condition. The gap spacings are based on minimum clearances in air for live point to earth as specified in various standards. For example, this clearance for 11 kV horn gap fuse is 180 mm as per German Standard VDE. In practice the horn gap length is generally of about 200 mm, 250 mm and 375 mm for 11 kV, 22 kV and 33 kV systems respectively.

② Noncurrent-limiting fuse. These are usually of the expulsion type. The arc is extinguished by the expulsion of gases produced by arcing initiating by the fusion of the element. The short circuit rupturing capacity of such fuses is higher because of the gases evolved inside the fiber tube which expel the arc products and quench the arc quickly.

(3) To know low voltage fuses. Figure 10.4 shows various types and application of low-voltage fuse.

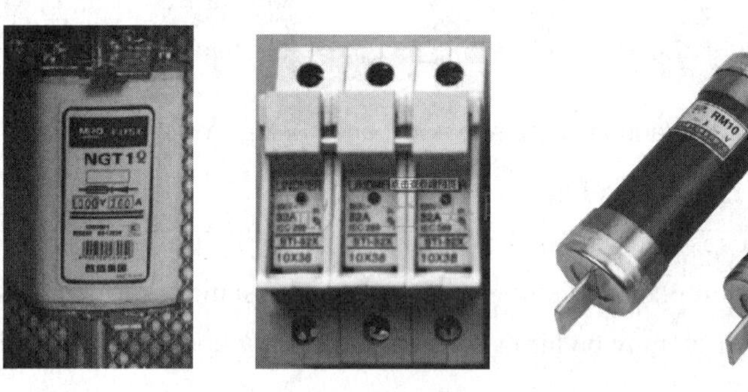

Figure 10.4 Various types and application of low-voltage fuse

These fuses are for voltage ratings up to 650 V. IS:3106—1966 covers procedure for the selection, installation and maintenance of these fuses. Low voltage fuses are of the following three types.

① Semi-enclosed type. It is very cheap rewirable fuse protection for circuits where a fairly low breaking-capacity is required. Copper tinned wire is generally used. The performance of these rewirable fuses is not guaranteed. Hence, whenever proper discrimination is required, these fuses are not recommended. Switch fuse units using these fuses give only coarse protection.

② Cartridge type fuse. Fuse links of this type are usually in completely enclosed cartridges, containing one or more fuse elements and filled with an arc-extinguishing medium, usually silica sand. The design of the fuse element is capable of considerable variations. IS:2208—1962 covers these fuses. The application of these fuses requires checking of voltage, current rating and class of fuses, i.e. quick action or delayed action, and adequacy of their breaking capacity. Apart from carrying the rated current continuously, fuses are generally designed to take a certain amount of overload for a pre-determined period of time before they operate. Installations which are designed to take very marginal overloads need protection even against small sustained overloads need protection even against small sustained overloads. In such cases an overload of only 25% should be sufficient to blow the fuse. Fuses which are classified having fusing factors (the ratio of the minimum current required for fuse operation to the rated current) between 1.25~1.5 and 1.5~1.75, respectively, can be used in circuits where the initial currents are higher or where the fuses are integrated in the system such that they do not blow on smaller over currents. IS:2208—1926 dose not provide such classification. The data regarding this can however, be procured from the manufacturer's catalogue (厂商目录). This is important to avoid unwarranted blowing of fuses.

These fuses are characterized by the property of "out-off" which means the ability to interrupt the flow of the fault current even before it reaches its maximum value. This property help in achieving very high rupturing capacities and restricts the flow of the fault current only up to a quarter of cycle. Indian standard 2208—1962 prescribes breaking capacity test of 46 kA for these fuses.

③ Miniature fuse. These are largely used for the protection of appliances and because of their relatively small breaking capacity. These are not employed in distribution system as much but are used in large numbers in the control apparatus associated will distribution systems, particularly consisting of electronic circuits. The principal advantage in such applications is the large of time-current characteristics at very low price. The types available vary from quick-acting and delay types in glass cartridges with no filler to miniature high breaking capacity types in porcelain bodies filled with silica sand.

2) Circuit breaker

A circuit breaker is a device which is not designed for frequent operation, but is capable of making and breaking all currents including fault currents up to its relative high rated breaking capacity. A wide variety of closing and tripping arrangements are available using relays with variable time delay and a number of operating mechanisms based upon solenoids, charged springs or pneumatic arrangements. The types of breakers used in a distribution system are: air break type, oil break type, vacuum type, and electronegative gas type (S_{F6} gas breaker). One great advantage of circuit breaker is their speedy operation, comparatively speaking, on a small overload and the considerable control of operating time under these conditions. Whereas a fuse characteristic is fixed in the design of the element, the operation of the breaker can be varied at will to suit variations in demand.

(1) To know the air and oil breaking circuit breaker, as shown in Figure 10.5.

The mode of action of all circuit breakers consists in the breaking of the fault current by the separation of a set of contacts. An arc is immediately established on separation of the contacts. The means by which this arc is extinguished enables distinction of different types of circuit breakers. When the contacts are closed and carrying current, the problem of temperature rise and millivolt drop, etc. is common to all forms of circuit breakers and contact force. Other features of the design also contribute but are of secondary importance.

For a voltage up to 11 kV, the plain break type of oil (bulk oil or minimum oil) circuit breaker is widely used to provide economical units of breaking capacities up to 250 MV · A. In this medium power range air break circuit breakers (ACBs) are used with breaking capacities up to 500 MV · A,11 kV. ACBs up to 7,000 A rating are in common use in LT (low tension) system. The arc is extinguished by the action of magnetic blow-out coils, which force it away from the contacts into an arc chute. The final break of contact and initiation of the arc is often accomplished by auxiliary arcing contacts which can be made of specially arc-resistant material. The arcing contacts are only required to carry the full current at the breaking instant since the main contacts carry the current all other times.

(a) Air breaking circuit breaker

(b) Oil breaking circuit breaker

Figure 10.5 Air and oil breaking circuit breakers

Although the air break circuit breakers are in general more expensive than the equivalent oil break types, their maintenance cost is generally lower, partly because of the need for changing oil in the latter. It is also considered necessary to install more sophisticated fire protection equipment with the oil break types.

The large range of control devices which can be used in conjunction with circuit breaker of both type, the case of fitting auxiliary switches, the ability to carry out switching at very high current and the rapid reclosure facility makes the circuit breaker the most versatile, although the most expensive, protective device.

(2) To know vacuum type circuit breakers, as shown in Figure 10.6.

(a) High voltage vacuum circuit breaker

(b) ZW10

Figure 10.6 Vacuum type circuit breakers

Vacuum interrupters up to 11 V, 22 V and 33 kV ratings are commonly used in European countries, their manufacture has been taken up in India. The technique of AC interruption in vacuum consists of separating a pair of current carrying contacts in a high vacuum environment of 10^{-5} Torr (1 Torr=133.322 Pa) or more. Unlike other arc interrupters, the current carriers in the arc are mainly metal ions from the contacts, since a negligible amount of gas is present in the arc plasma. The memory of the arc is shortened by the condensation of these mental ions and vapours on to the contacts and shields and the effectiveness of this determines the efficiency of the interruption process at zero current. Vacuum circuit breakers offer the following advantages.

① Long life with minimum maintenance.

② Completely enclosed and sealed construction.

③ Extremely short and consistent arcing and total break times.

④ Suitability for very fast automatic reclosure.

⑤ No fire risk.

⑥ No noise and no emission of gas or air during operation.

⑦ At the end of its life, a bottle can be quickly removed, discarded and replaced by another spare bottle having an extremely long shelf life.

(3) To know the SF_6 circuit breaker, as shown in Figure 10.7.

Figure 10.7 SF_6 circuit breaker

The superior arc-quenching ability of SF_6 gas can be attributed to the fact that it is electronegative, which means that its molecules rapidly absorb the free electrons in the arc path between the breaker contacts to form negatively charged ions which are ineffective as current carriers. The electron trapping action results in a rapid build up of the insulation strength after zero current. For effective arc extinction it is, however, necessary to force the gas into the arc, but the properties of SF_6 are such that the gas flow velocity does not need to be high as in the air blast circuit breakers.

Among the advantages claimed for the SF_6 circuit breakers are:

① The low gas velocity and pressures employed minimize any tendency towards current chopping, thereby the capacitive currents can be interrupted without restriking.

② The closed circuit gas cycle coupled with low gas velocity gives quiet operation as there is no exhaust to atmosphere as in air blast breakers.

③ The closed gas circuit keeps the interior dry so that there is no moisture problem.

④ The arc extinguishing properties of SF_6 gas result in very short arcing times so that contact erosion is very little. The contacts can be run at higher temperatures without deterioration.

⑤ There are no carbon deposits so that tracking and insulation breakdown is eliminated.

⑥ Electrical clearances can be reduced due to the insulating properties of the gas.

⑦ As the circuit breaker is totally enclosed and sealed form atmosphere, it is particularly suitable for use in coalmines or in any industry where explosion hazard exists. Breakers in lower voltage range are used at a station where already higher voltage range (220 kV and above). SF_6 breakers have been installed for reasons of economy. Distribution breakers of this type may not be economical to use in isolation except in special circumstances such as in mines, etc., Recently, the manufacture of these breakers has been taken up in India by Voltas, Bombay.

Extension Materials

Introduction to addition protection of equipment

Individual items of equipment frequently need special protective devices because of their inherent operating characteristics. These are, of course, beyond the distribution stage, but are of interest in the design of the distribution system because of their possible effects on discrimination. Two examples illustrating this problem are given below.

1. Protection of motors

Two distinct principles are employed: operation by current and tripping by inbuilt thermal protective sensors in the motor.

1) Current operated devices

These comprise either a trip circuit. Incorporated in the contactor in series with the circuit current, the trip level being adjusted to a precalibrated setting or a separate relay device operated form series coils or current transformers. The time-operating current characteristic of these devices is designed to pass the initial motor starting surge (浪涌) but to trip as soon as the motor stalls.

The protective devices of this type have the advantage that they do not require additional connections to the motor, and if the current response of the device matches the heating characteristics of the motor, it can provide protection when dangerous changes occur. The disadvantage of this type of protection is that the heating characteristics differ for different motors.

2) Temperature sensing protective devices

These devices are embedded in the motor and operated a control circuit which can switch off the current when the device reaches a predetermined temperature. They have the advantage that they immediately respond to excessive temperature rise in any crucial part of the motor windings, bearings or any other vital position in which they are placed. In addition, since they are not the current, they permit any rough usage or overloading which does not produce a destructive temperature rise. The barium titanate thermistor of the positive temperature coefficient type is probably the most convenient and responsive of these devices. Also, if it is true that the criterion of damage by misuse of motor is its temperature rise at crucial points, it would appear that this method has considerable advantages. Set against these advantages is the necessity for an additional control circuit, usually a solid state control module containing resistors, capacitors, Zener diodes and transistors which sense the changes in the thermistor and cause a thyristor to perform the switching operation.

2. Fuse switch combinations

The accurate current limiting operation and high breaking capacity of fuses has resulted in a number of ingenious combinations of switches and fuse to get an economical unit combining the best performance of each. These combinations find a large number of such applications. Three examples are given below to illustrate their cases.

One example is the limitation by series fuses of short circuit energy in contactors, which are designed only for breaking overload currents. This is a special case of backup protection, as shown in Figure 10.8.

Another example is the use of fuses across the first break to open a load breaking switch. Here the "limiter" type of fuse finds application, since it is only required to operate on heavy short circuit and does not normally carry current.

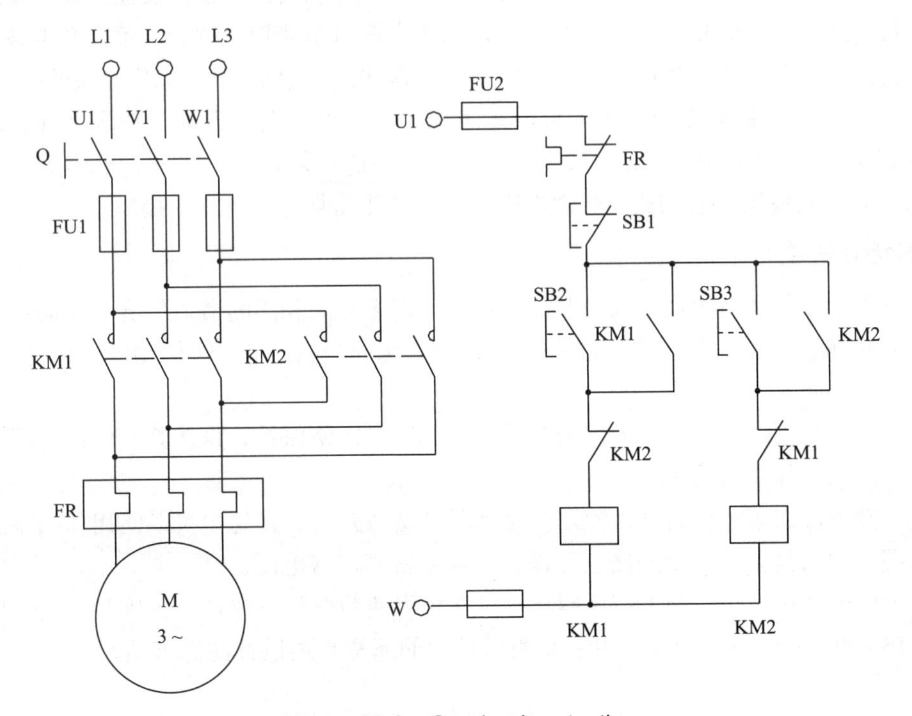

Figure 10.8 Overload protection

A third example is the use of fuses across trip coils to produce rapid, precise and decisive tripping of the breaker at a limiting load current. IS:4047 and 4064 discuss all the important design considerations for switch fuse units.

设备附加保护装置介绍

个别设备由于其固有的工作特性，经常需要特殊的保护装置。这些当然超出了分配范围，但由于它们可能对系统辨别能力产生影响，因此在分配系统的设计中引起了人们的兴趣。下面给出两个例子说明此问题。

1. 电动机保护

采用两种截然不同的原理：通过电流操作和通过电动机中内置的热保护传感器断开。

1）电流动作装置

这些包括跳闸电路。在电路中与接触器串联，调整动作电流到预设值，或者由串联线圈或电流互感器操作独立的继电器装置。这些装置的实时电流特性设计是允许通过电动机的初始浪涌电流，一旦电动机失速就会跳闸。

这种类型的保护装置的优点在于，它们不需要与电动机的额外连接，并且如果该装置的电流响应与电动机的加热特性相匹配，则可以在发生危险时提供保护。这种保护的缺点是加热特性随不同的电动机而不同。

2）温度感应保护装置

这些设备嵌入电动机中，并运行控制电路，当设备达到预定温度时，该控制电路可以切断电流。它们的优点是，可以立即对电动机绕组、轴承或它们所处的任何其他重要位置的温度过高立即做出响应。此外，由于它们不是电流，因此允许粗暴使用或过载，不会产生破坏性的温度上升。正温度系数类型的钛酸钡热敏电阻可能是这些设备中最方便、响应最快的一种。同样，如果确实因误用电动机而造成损害的标准是其在关键点处的温度升高，那么这种方法似乎具有很大的优势。需要附加控制电路来优化这些特点，通常是包含电阻器、电容器、二极管和晶体管的静态控制模块，它们感测热敏电阻的变化并使晶闸管执行开关操作。

2. 熔丝开关组合

精确的限流操作和熔断器的高分断能力已产生许多开关和熔断器的组合，从而获得了一种结合了各自的最佳性能经济单元。这些组合得到了大量的应用。下面给出了三个例子来说明它们的情况。

第一个例子是电路中用接触器与熔断器串联来进行短路保护，该熔断器仅设计用于断开过载电流。这是后备保护的特例。

另一个例子是在第一次断开时使用熔丝来断开负载开关。在这里可以使用“限制器”类型的熔丝，因为它仅需要在严重短路下工作，并且通常不承载电流。

第三个例子是在跳闸线圈上使用熔丝，以在极限负载电流下快速、精确和决定性地使断路器跳闸。IS：4047 和 4064 介绍了开关熔断器单元的所有重要设计注意事项。

Self-Test

(1) Choose the best answer into the blank.

① Protection is provide to ______________ some hazards in power system.

A. remove B. minimize C. increase

② All fuses having a breaking capacity of ______________ can be referred to as high rupturing capacity (HRC) fuses.

A.16 kA B. over16 kA C. Under 16 kA

③ A cartridge type fuses is ____________________.

A. a rewirable B. a high voltage fuse C. a low voltage fuse

④ Low voltage fuses are for voltage ratings up to __________________.

A.132 kV B.650 kV C.650 V

⑤ ACBs up to 7,000 A rating are in common use in _______________ system.

A.LT B.HT C. any

⑥ Temperature sensing protective devices operate a control circuit which can ______________ the current when the device reaches a predetermined temperature.

A. switch off B. witch on C. operate

⑦ A protective relay must operate at the __________________ speed and must be reliable.

A. required B. allow C. quick

⑧ Current transformers are designed to provide a standard secondary current output of ______

A. 0 A or 5 A B. 1 A or 10 A C. 1 A or 5 A

(2) Complete the sentences.

① Two types of high voltage fuse are _________________ and _________________.

② Three types of low voltage fuse are ___________________________, ____________________and ________________________.

③ The types of breaks used in a distribution system are: ________________ , _________________, _______________and _____________________________________.

④ One great advantage of circuit breakers is _____________________________________.

⑤ Two protective devices of motors are _________________________ and ___________________.

New Words and Phrases

anticipate *vt.* 预感；预见；预料

hazard *vt.* 冒险；使遭受危险 *n.* 危险；冒险的事

shock *n.* 休克；震惊；震动；打击

vt. 使休克；使震惊；使震动；使受电击

discrimination *n.* 歧视；辨别，区别；辨别力，识别力；不公平的待遇

branch *n.* 树枝；分支；部门，分科；支流

vi. 分支形成；分支扩张；扩大某人的兴趣，业务或活动范围；[计]下分支的指令

vt. 使分支；使分叉；用枝形叶脉刺绣花纹装饰

fault *n.* 缺点，缺陷；过错，责任；[电]故障

vt. 挑剔，找……的缺点；批评；做错，在……中出错

vi. 找错误，挑剔

arc *n.* 综合症状；弧（度）；天穹；电弧，弧光。

vi. 形成拱状物；循弧线行进

HRC（high rupturing capacity） 高切断能力

Silica *n.* 硅石，二氧化硅

porcelain *n.* 瓷器，瓷

quench *vt.* 解（渴）；终止（某事物）；(用水）扑灭（火焰等）

arc-extinguishing 灭弧

copper tinned wire 镀锡铜线

ACB（air break circuit breaker） 空气断路器

LT (low tension) 低电压

magnetic blow-out coil 电磁灭弧线圈

Barium *n.* 钡

Titanate *n.* 钛酸盐

thyristor *n.* 晶闸管

Zener diode 齐纳二极管

CHAPTER 5

Mechatronics Technology

Objectives:

(1) Trace the origin of mechatronics.

(2) Understand the key elements of mechatronics systems.

(3) Relate to everyday examples of mechatronics systems.

(4) Appreciate how mechatronics integrates knowledge from different disciplines in order to realize engineering and consumer products that are useful in everyday life.

Unit 11

Mechatronics

With the technical advance, the use of machinery has superseded manual labor. Figure 11.1 shows the application of mechatronics technology in industrial production.

Figure 11.1 Electromechanical control system

1. Historical perspective

Advances in microchip and computer technology have bridged the gap between traditional electronic, control and mechanical engineering. Mechatronics responds to industry's increasing demand for engineers who are able to work across the discipline boundaries of electronic, control and mechanical engineering to identify and use the proper combination of technologies for optimum solutions to today's increasingly challenging engineering problems. All around us, we can find mechatronic products. Mechatronics covers a wide range of application areas including consumer product design, instrumentation, manufacturing methods, motion control systems, computer integration, process and device control, integration of functionality with embedded microprocessor control, and the design of machines, devices and systems possessing a degree of computer-based intelligence. Robotic manipulators, aircraft simulators, electronic traction control systems, adaptive suspensions, landing gears, air conditioners under fuzzy logic control, automated diagnostic systems, micro electromechanical systems (MEMS), consumer products such as VCRs, and driver-less vehicles are all examples of mechatronic systems.

The genesis of mechatronics is the interdisciplinary area relating to mechanical engineering, electrical and electronic engineering, and computer science. This technology has produced many new products and provided powerful ways of improving the efficiency of the products we use in our daily life. Currently, there is no doubt about the importance of mechatronics as an area in science and technology. However, it seems that mechatronics is not clearly understood; it appears that some people think that mechatronics is an aspect of science and technology which deals with a system that includes mechanisms, electronics, computers, sensors, actuators and so on. It seems that most people define mechatronics by merely considering what components are included in the system and/or how the mechanical functions are realized by computer software. Such a definition gives the impression that it is just a collection of existing aspects of science and technology such as actuators, electronics, mechanisms, control engineering, computer technology, artificial intelligence, micro-machine and so on, and has no original content as a technology. There are currently several mechatronics textbooks, most of which merely summarize the subject picked up from existing technologies. This structure also gives people the impression that mechatronics has no unique technology. The definition that mechatronics is simply the combination of different technologies is no longer sufficient to explain mechatronics. Mechatronics solves technological problems using interdisciplinary knowledge consisting of mechanical engineering, electronics, and computer technology. To solve these problems, traditional engineers used knowledge provided only in one of these areas (for example, a mechanical engineer uses some mechanical engineering methodologies to solve the problem at hand). Later, due to the increase in the difficulty of the problems and the advent of more advanced products, researchers and engineers were required to find novel solutions for them in their research and development. This motivated them to search for different knowledge areas and technologies to develop a new product (for example, mechanical engineers tried to introduce electronics to solve mechanical problems). The development of the microprocessor also contributed to encouraging the motivation. Consequently, they could consider the solution to the problems with wider views and more efficient tools; this resulted in obtaining new products based on the integration of interdisciplinary technologies. Mechatronics gained legitimacy in academic circles with the publication of the first refereed journal: *IEEE/ASME Transactions on Mechatronics*. In it, the authors worked tenaciously to define mechatronics. Finally they coined the following: The synergistic combination of precision mechanical engineering, electronic control and systems thinking in the design of products and manufacturing processes. This definition supports the fact that mechatronics relates to the design of systems, devices and products aimed at achieving an optimal balance between basic mechanical structure and its overall control.

2. Key elements of a mechatronic system

It can be seen from the history of mechatronics that the integration of the different technologies to obtain the best solution to a given technological problem is considered to be the essence of the discipline. There are at least two dozen definitions of mechatronics in the literature but most of them hinge around the "integration of mechanical, electronic, and control engineering, and information technology to obtain the best solution to a given technological problem, which is the realization of a product"; we follow this definition. Figure 11.2 shows the main components of a mechatronic system. As can be seen, the key element of mechatronics are electronics, digital control, sensors and actuators, and information technology, all integrated in such a way as to produce a real product that is of practical use to people.

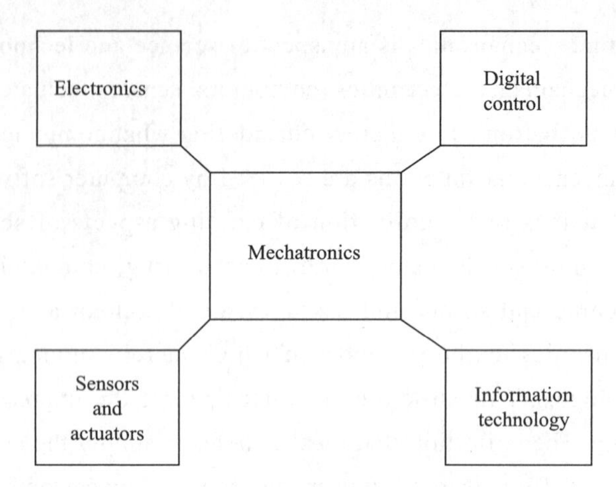

Figure 11.2 Main components of a mechatronic system

1) Electronics

Semiconductor devices, such as diodes and transistors, have changed our lives since the 1950s. In practice, the two most commonly used semiconductors are germanium and silicon (the latter being most abundant and cost-effective).

2) Digital control

A transfer function defines the relationship between the inputs to a system and its outputs. The transfer function is typically written in the frequency (or s) domain, rather than the time domain. The Laplace transform is used to map the time domain representation into the frequency domain representation.

3) Sensors and actuators

Sensors are elements for monitoring the performance of machines and processes. The common classification of sensors is: distance, movement, proximity, stress/strain/force, and temperature. There are many commercially available sensors but we have picked the ones that are frequently used in mechatronic applications. Often, the conditioned signal output from a sensor is transformed into a digital form for display on a computer or other display units. The apparatus for manipulating the sensor output into a digital form for display is referred to as a measuring instrument (see Figure 11.3 for a typical computer-based measuring system).

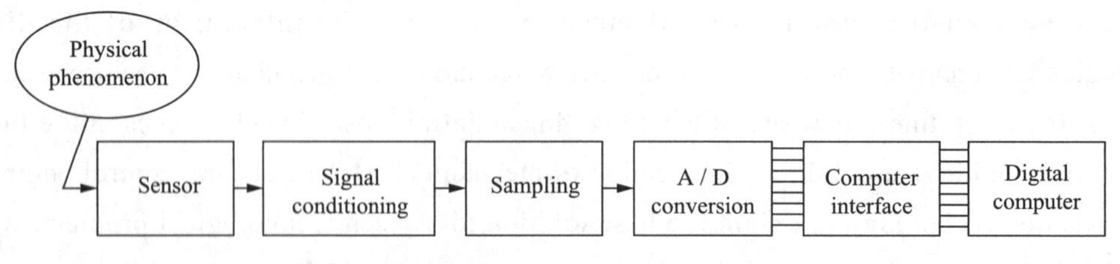

Figure 11.3 A typical computer-based measuring system

While a sensor is a device that can convert mechanical energy to electrical energy, an electrical actuator, on the other hand, is a device that can convert electrical energy to mechanical energy. All actuators are transducers (as they convert one form of energy into another form). Some sensors are transducers (e.g. mechanical actuators), but not all. Actuators are used to produce motion or action, such as linear motion or angular motions. Some of the important electrical actuators used in mechatronic systems include solenoids, relays, electric motors (stepper, permanent magnet, etc.). These actuators are instrumental in moving physical objects in mechatronic systems.

Mechanical actuators are transducers that convert mechanical energy into electrical energy. Some of the important mechanical actuators used in mechatronic systems include hydraulic cylinders and pneumatic cylinders.

4) Information technology

Signals to and from a computer and its peripheral devices are often communicated through the computer's serial and parallel ports. The parallel port is capable of sending (12 bits per clock cycle) and receiving data (up to 9 bits per clock cycle). The port consists of four control lines, five status lines, and eight data lines. Parallel port protocols were recently standardized under the IEEE 1284 standard. These new products define five modes of operation such as:

(1) Compatibility mode.

(2) Nibble mode.

(3) Byte mode.

(4) EPP (enhanced parallel port) mode.

(5) ECP (extended capabilities ports) mode.

3. Some examples of mechatronic systems

Today, mechatronic systems are commonly found in homes, offices, schools, shops, and of course, in industrial applications. Common mechatronic systems include:

(1) Domestic appliances, such as fridges and freezers, microwave ovens, washing machines, vacuum cleaners, dishwashers, cookers, timers, mixers, blenders, stereos, televisions, telephones, lawn mowers, digital cameras, videos and CD players, camcorders, and many other similar modern devices.

(2) Domestic systems, such as air conditioning units, security systems, automatic gate control systems.

(3) Office equipment, such as laser printers, hard drive positioning systems, liquid crystal displays, tape drives, scanners, photocopiers, fax machines, as well as other computer peripherals.

(4) Retail equipment, such as automatic labeling systems, bar-coding machines, and tills found in supermarkets.

(5) Banking systems, such as cash registers, and automatic teller machines.

(6) Manufacturing equipment, such as numerically controlled (NC) tools, pick-and-place robots, welding robots, automated guided vehicles (AGVs), and other industrial robots.

(7) Aviation systems, such as cockpit controls and instrumentation, flight control actuators, landing gear systems, and other aircraft subsystems.

Extension Materials

Application of mechatronics technology

The success of industries in manufacturing and selling goods in a world market increasingly depends upon an ability to integrate electronics and computing technologies into a wide range of primarily mechanical products and processes. The performance of many current products—cars. washing machines, robots or machine tools—and their manufacture depend on the capacity of industry to exploit developments in technology and introduce them at the design stag into both products and manufacturing processes. The results is systems which are cheaper, simpler, more reliable and with a greater flexibility of operation than their predecessors. In this highly competitive situation, the old divisions between electronic and mechanical engineering are increasingly being replaced by the integrated and interdisciplinary approach to engineering design referred to mechatronics.

In a highly competitive environment, only those new products and processes in which an effective combination of electronics and mechanical engineering has been achieved are likely to be successful. In general, the most likely cause of a failure to achieve this objective is an inhibition, on the application of electronics. In most innovative products and processes the mechanical hardware is that which first seizes the imagination, but the best realization usually depends on a consideration of the necessary electronics, control engineering and computing from the earliest stages of the design process. The integration across traditional boundaries that this implies and requires lies at the heart of a mechatronic approach to engineering design and is the key to understanding the developments that are taking place.

To be successful, a mechatronic approach needs to be established from the very earliest stages of the conceptual design process, where options can be kept open before the form of embodiment is determined. In this way the design engineer, and especially the mechanical design engineer, can avoid going too soon down familiar and perhaps less productive paths.

Where full attention has been given to market trends, the adoption of an integrated mechatronic approach to design has led to a revival in areas such as high speed textile equipment, metrology and measurement systems, and special purpose equipment such as that required for the automatic testing of integrated circuits. In most cases, the revival or new growth is brought about by the enhancement of process capability achieved by the integration of electronics, often in the form of an embedded microprocessor, with the basic mechanical system.

This demand for increased flexibility in the manufacturing process has led to the development of the concept of flexible manufacturing systems (FMSs) in which a number of elements such as computer numerically controlled machine tools, robots and automatically guided vehicles (AGVs) are linked together for the manufacture of a group of products. Communication between the individual elements of the system is achieved by means of local area networks (LANs).

Within products, the diversity and opportunity offered by a mechatronic approach to engineering design is to date largely unrealized. End user products are substantial revenue earners and it is possible here to distinguish between existing products offering enhanced capabilities and completely new product areas that would not have existed without a mechatronic design approach having been adopted from the outset.

In the first category the following are illustrative from many examples.

1. Automotive engines and transmissions

Engine and driveline management systems leading to reduced emissions, improved fuel speeds, and selectable gear characteristics.

2. Power tools

Modern power tools such as drills offer a range of features including speed and torque control, reversing drives and controlled acceleration.

Examples in the second category include the following.

1. Modular robotics

Conventional industrial robots are often limited in their operation by their geometry. By providing a range of structural components and actuators together with a central controller a modular robotics system has been made available, allowing users to assemble robot structures directly suited to their needs.

2. Video and compact disc players

Video and compact disc player involve complex laser tracking systems to read the digitally encoded signal carried by the disc. This control is achieved by means of a microprocessor based system that also provides features such as multiple track selection, scanning and preview.

A common factor in consumer mechatronics as exemplified by the above is the continuous improvement in capability achieved against a constant or reducing real cost to the end user. The capability of a mechatronic system, based as it often is on inexpensive components or modules, also provides a means to execute bespoke solutions to special problems.

In engineering design, a mechatronics approach represents and requires the integration of a wide rang of material and information aimed at providing systems which are more flexible and of higher performance than their predecessors, and which incorporate a wide range of features. Thus for full benefit and effect, mechatronics must be a feature of both the conceptual and embodiment phases of the design process.

In manufacturing, users are demanding a much higher degree of control of both the overall process and its components. This requires knowledge both of the capabilities of these components and of the means by which they are integrated within the complete system.

机电一体化

在国际市场中，制造业和工业产品销售业绩取得的成绩，越来越依靠电子技术和计算机技术与传统机械制造和机械产品的广泛结合。目前这些产品（如汽车、洗衣机、机器人和机床）的性能及制造，主要依靠工业对新技术的开发，并把它们应用于产品及制造过程的各个设计阶段。结果使整个工业系统产生了比过去更便宜、更简单、更加可靠、功能更强的制造技术，这种激烈的竞争，导致原来的电子工程和机械工程的区别逐渐被各学科工程设计之间互相结合与互相渗透所取代，产生了机电一体化，又称机械电子学。

在这个激烈的竞争环境里，获得成功的产品和技术都是那些把电子和机械有效结合的产品，而没有成功的主要原因也是由于没有应用电子技术。一般产品机器制造工业的革新，往往从机械的硬件设计开始，但要想实现设想，则从设计过程的最初阶段就要充分考虑电子技术、控制工程及计算机技术。从机械电子学研究到工程设计，其关键是穿越机械学和电子学的隐分界，把它们结合起来，这是理解当今发生的这场变革的关键。

要想成功，在设计的初期就需要建立机电一体化的研究概念，这时具体方案还没有形成，所以还有选择的余地。这样，设计工程师特别是机械设计工程师就能避免过快地做出决定而落入俗套，降低生产率。

充分研究市场的发展趋势，我们会发现采用一项机电一体化的设计方案，会导致某个领域的复兴。例如高速纺织机、计量与测量系统以及像集成电路自动测试装备那样的专用设备。在很多情况下，新兴领域的产生和复兴往往由内嵌式微处理器形成的电子学与基本机械系统的综合而引起，增强了加工能力。

制造过程灵活性的要求导致了柔性操作系统概念的产生，在这套系统中，许多元件，例如计算机数控机床、机器人和自动导引小车等联系在一起共同进行生产，它们之间的信息交流通过局域网实现。

迄今为止的产品中，大都没有实现机电一体化为工程设计提供的多样性和机遇。销售给用户的最终产品才是我们收入的本质来源，这可能才是迄今为止才开始应用的全新的机电一体化产品与提供增强功能的传统产品的重要区别。

下面的例子可以说明那些传统产品。

1. 自动控制发动机和自动控制变速器

发动机及其传动装置的发展趋向于降低辐射、节省燃料以及通过防止低速时过多的燃料流入和使用可以调节齿轮来避免误驱动等。

2. 动力驱动工具

现代动力驱动工具，例如钻头能够提供多种功能，包括速度和扭矩控制、反向驱动及加速度控制。

全新的机电一体化产品的例子如下：

1. 标准组件组装的机器人

传统的工业机器人由于结构问题往往受到许多限制。使用一些结构零件与驱动装置，再加上中央处理器，就可以制成由标准组件组装的机器人系统，这样用户就可以直接组装满足他们自己需要的各种机器人。

2. 视频和 CD 播放器

视频和 CD 播放器装有复杂的激光磁头，可以读出磁盘上的数字信息。借助微处理器，控制系统可以提供多磁道选择、扫描、预览等多种功能。

上面的例子表明，使用机电一体化技术产品的目的是不断改进消费品，而不是持续降低消费品的价格。机电一体化产品提供了解决各种特殊问题的理想方式，使用的是低成本元件或标准件。

在工件设计中，机电一体化代表并要求有广泛的材料和信息的综合，其产品功能更强，使用更灵活，特征更广泛。只有在设计概念和具体的设计阶段都应用了机电一体化技术，才能得到较好的收益和效率。

在工程制造中，用户要求充分控制整个制造过程及其各个元件。这既需要了解各类电子元件的知识，也需要了解将它们集成在整个系统中的方法。

Self-Test

Answer the following questions.

(1) What do you understand by the term “mechatronics”?

(2) What are the key elements of mechatronics?

(3) Is mechatronics the same as electronic engineering plus mechanical engineering?

(4) Is mechatronics as established as electronic or mechanical engineering?

(5) List some mechatronic systems that you see everyday.

New Words and Phrases

mechatronics *n.* 机电一体化，机械电子学

exploit *v.* 开发，利用，使用，发挥

electronics *n.* 电子学，电子仪器

predecessor *n.* 前辈，前任，被取代的事物

inhibition *n.* 抑制，制止，阻止，阻碍，延缓

innovative *adj.* 革新的，创新的，富有革新精神的

embodiment *n.* 具体化，具体装置

revival *n.* 复兴，恢复，再生，再留行

textile *adj.* 纺织的

metrology *n.* 计量学，测量学

enhancement *n.* 增强，加强，提高，放大

substantial *adj.* 物质的，本质的，显著的

revenue *n.* 收入，收益

outset *n.* 开端，开始，最初

illustrative *adj.* 说明的，解说的，例证的

emission *n.* 发射，放出，排出物

misuse *n.* 错用，误用，滥用

reverse *v.* 颠倒，改变方向，转换

modular *adj.* 制成标准组件的，预制的，组合的

track *v.* 跟踪，沿轨道行驶

module *n.* 模块，组件

bespoke *adj.* 专做订货的

execute *v.* 实行，完成，编制，操纵

incorporate *v.* 使结合，联合，合并，组成公司

Unit ⑫

Machine Tools

1. Machine tools

Most of mechanical operations are commonly performed on basic machine tools:

1) The drill press (see Figure 12.1)

Figure 12.1 The drill presses

Drilling is performed with a rotating tool called a drill. Most drilling in metal is done with a twist drill. The machine used for drilling is called a drill press. Operations, such as reaming and tapping, are also classified as drilling. Reaming consists of removing a small amount of metal from a hole already drilled. Tapping is the process of cutting a thread inside a hole so that a cap screw or bolt may be threaded into it.

2) The lathe (see Figure 12.2)

Figure 12.2 The lathes

The lathe is commonly called the father of the entire machine tool family. For turning operations, the lathe uses a single-point cutting tool which removes metal as it travel past the revolving workpiece. Turning operations are required to make many different cylindrical shapes, such as axes, gear blanks, pulleys, and threaded shaft.

Boring operations are performed to enlarge, finish and accurately locate holes.

3) The shaper or planer (see Figure 12.3 and Figure 12.4)

Figure 12.3 The shaper

Figure 12.4 The planer

Shaping and planning produce flat surfaces with a single-point cutting tool. In shaping, the cutting tool on a shaper reciprocates or moves back and forth while the work is fed automatically towards the tool. In planning, the workpiece is attached to a worktable that reciprocates past the cutting tool. The cutting tool is automatically fed into the workpiece a small amount on each stroke.

4) The milling machine or borer (see Figure 12.5 and Figure 12.6)

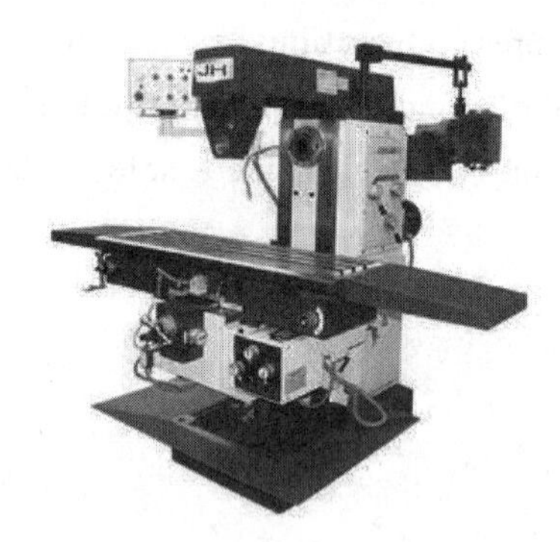

Figure 12.5 The milling machine

Figure 12.6 The borer

Milling removes metal with a revolving, multiple cutting edge tool called milling cutter. Milling cutters are made in many styles and sizes. Some have as few as two cutting edges and others have 30 or more. Milling can produce flat or angled surfaces, groves, gear teeth, and other profile. Depending on the shape of the cutters being used.

Boring operations are performed to enlarge, finish and accurately locate holes.

5) The grinder (see Figure 12.7)

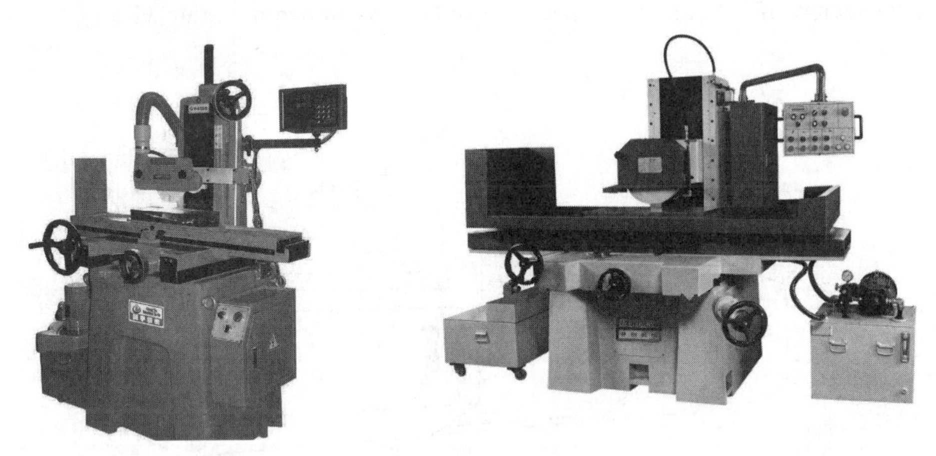

Figure 12.7 The grinders

Grinding makes use of abrasive particles to do the cutting. Grinding operations may be classified as precision or nonprecision, depending on the purpose. Precision grinding is concerned with grinding to close tolerances and very smooth finish. Nonprecision grinding involves the removal of metal where accuracy is not important.

2. Numerical control machines

The modern manufacturing industry has high mechanical processing scope and processing accuracy. Traditional processing machines can no longer meet the production requirements. Thus, the numerical control machines evolved. In recent years, numerical control technology has been developing rapidly. It is used widely in manufacturing industry. It is of great importance to learn the fundamental knowledge concerning numerical control machines.

The numerical control machine can be defined as a machine with a program control system. The numerical control machines may be classified into common numerical control machines and numerical control machines of processing centers according to processing methods and technics purposes.

1) Common numerical control machines

Commonly used numerical control machines include the drilling machine, the lathe and the milling machine.

These machines have three purposes. First, they hold the workpiece or the part to be cut. Second, they hold the cutting tools. Third, they regulate the cutting speed and the feeding movement between the tool and the workpiece.

2) Numerical control lathes

The lathe is one of the most productive machines. It is an efficient tool when producing round parts. Most of the lathes are programmed on two axes.

(1) The X axis controls the cross movement (in or out) of the cutter.

(2) The Z axis controls the carriage approach or depart from the headstock.

The outside view of CK6136 numerical control lathe as shown in Figure 12.8.

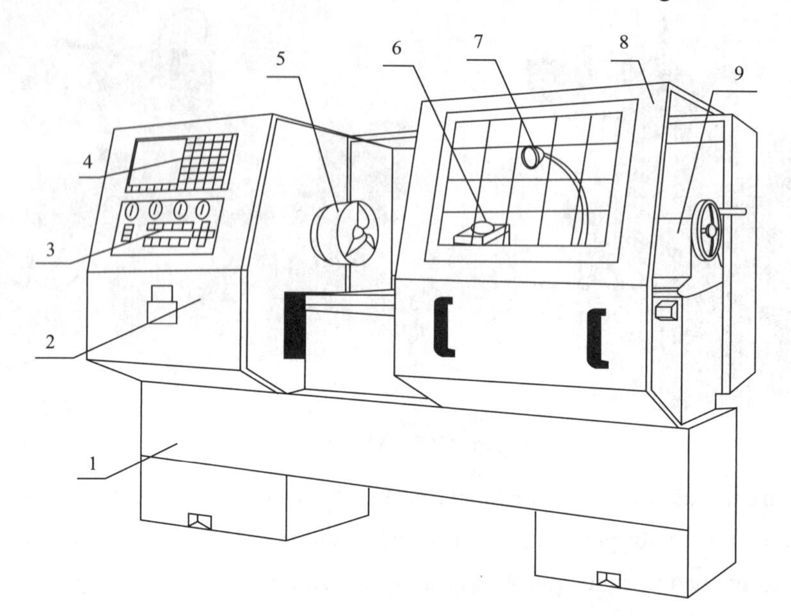

Figure 12.8 The outside view of CK6136 numerical control lathe

1-The lathe bed; 2-The headstock; 3-The operation panel of the lathe; 4-The operation panel of numerical control system; 5-The spiral chuck; 6-The swiveling tool-holder; 7-The illuminating lamp; 8-The protective door; 9-The tailstock

A lathe is a machine used primarily for producing surfaces of revolution and flat edges. There are several types of lathes such as horizontal lathes, vertical lathes, turret lathes, automatic lathes and lathes for special purposes. The lathe is able to carry out various machining operations.

In the course of revolving the workpiece, a cutter moves in the direction parallel to the principal axis or at an angle to the principal axis, then it cuts off the metal from the surface. The movement of the cutter is called the feed. The cutter is installed in the tool post on the carriage.

The carriage is the tool that controls the cutter's moving direction according to requirements.

The largest component of a lathe is the lathe bed. At either end of the lathe bed are the headstock and the tailstock. On the upper part of the lathe bed is the special orbit for the carriage and the tailstock to move.

The two lathe centers are installed in two axes holes. One is in the headstock while the other is in the hole of tailstock spindle.

The clamp is used to fix the workpiece in order to prevent it from wobbling in the process of revolution.

When processing workpieces of different materials and diameters, the lathe should run at different speeds. The speed-change gearbox in the headstock enables the lathe to run at different speeds. Before revolving the workpiece in the lathe, you should align the lathe centers. That is to say, both centers of the axes are supposed to be on the line. Not all the workpieces should be fixed between the lathe centers. When turning a short workpiece, the dead center is not needed. You can just clamp it properly in the middle of the headstock.

3) Milling ways of numerical control milling machines

The operations on the milling machine are circular milling and head face milling. The most convenient and easiest milling way is single-piece milling. Here is the process: The operator puts the workpiece in the clamp or on the worktable, fixing it at the proper location. After that, the operator should make the worktable move forward rapidly in order that the workpiece will be taken to the right place and be milled. After being milled, the workpiece returns to the original place. There, you can unload it and substitute a new one. Then the cycle can be repeated.

In the following part, you will be introduced the head face milling in details.

(1) The preparations of the milling:

① Select the proper head face milling cutter, then fix it on the axis.

② Fix the workpiece in the parallel jaw, then, fix the milling machine on the worktable with the parallel jaw.

③ Adjust and choose the proper milling speed and feed amount.

(2) Cutting:

① Move the raising platform so that the workpiece is focused on the cutter. Next, move the planker in so as to make the workpiece slightly apart from the cutter surface. Then move the worktable in order that the workpiece is away from the cutter. Adjust the cross-feed micrometer index to zero.

② Turn on the power, then, examine the cutter's revolution.

③ According to the cutting depth of the rough processing, turn the cross-feed mechanism.

④ Make the workpiece feed towards the cutter by hand, and check the cutter's action. If the cutting action is correct, make the worktable feed. Be sure to use enough cutting fluid.

⑤ Before you turn off the automatic feed, make sure the workpiece is completely apart from the cutter. If you ignore this, the surface will not be completely flat.

⑥ When cutting is completed, return the workpiece to its original position. Then, measure the length of the workpiece. If needed, you may take more cuts.

⑦ When the accurate processing is done, unload the apparatus and return the cutters and the clip apparatus to original position.

Extension Materials

CNC Machines

1. What is CNC

Early machine tools were operated by craftsmen who decided many variables such as speeds, feeds, and depth of cut, etc.

With the development of science and technology, a new term, numerical control (NC) appeared. Controlling a machine tool using a punched tape or stored program is known as numerical control. When numerical control is performed under computer supervision, it is called computer numerical control (CNC).

Computers are the control units of CNC machines, they are built in or linked to the machines via communications channels. CNC means the operation of machine tools and other processing machines by a series of commands of the tool to the work-piece. An organized list of commands constitutes a CNC program. The program may be used repeatedly to obtain identical results. When the job changes, we can only change the program of instructions. The capability to change the program for each new job gives CNC its flexibility.

2. CNC machine tools

CNC technology is being used on all types of machine tools (see Figure 12.9), from the simplest to the most complex, including the single-spindle drilling machine, lathe and turning center, vertical or horizontal-spindle machine center and milling machine, punches, electrical discharge machines (EDM), flame cutters, grinders, testing and inspection equipment, etc.

Figure 12.9 CNC machine tools

3. Three units of CNC

In general, any CNC machine tool consists of the followings units:

1) CNC device

As with all computers, the CNC device works on a binary principle using only two characters 1 and 0. Series of ones and zeros which is so-called machine language is the only language the computer understands. When creating the program, operators use a list of codes and keys to express the job information.

After inputting the program, CNC device will first compile it into machine language with special built-in software, then the computer will calculate all the points and send out the control commands (electronic pulse signals) to serve drive system

2) Serve drive system

This system includes drives and serve motors, which is the bridge connecting CNC device and machine. Drives transform pulse signals sent out from CNC device into motor signals, which tell the serve motor to rotate with each pulse. The rotation of the serve motor in turn rotates the lead screw which drives the linear axis. With each step of movement, the measure device sends a signal back to CNC device, which compares the current position of the drive axis with the programmed position.

3) Machine

The mechanism of numerical control machine tools consists of two parts. Except the common parts, such as the host movement part, the feed movement part (worktable, tool rest), the auxiliary body (hydraulic, cooling and lubrication apparatus) and the strut part (lathe bed, column) and so on, it also includes some special parts, like magazines and automatic tool changers.

Most of the time, several different cutting tools are used to produce a part. The tools must be replaced quickly for the next machining operation. For this reason, the majority of CNC machine tools are equipped with automatic tool changers. They allow tool changing without the intervention of the operator. Typically, an automatic tool changer grips the tool in the spindle, pulls it out , and replaces it with another tool. On most machines with automatic tool changers, the magazine can rotate in either direction, forward or reverse.

数控机床

1. 什么是计算机数控机床

老式机床通常是由工人操作并由他们决定机床速度、进给量、切削深度等。

随着科学技术的发展，一个新术语——数字控制（NC）诞生了。数字控制就是利用穿孔纸带或存储的程序来控制机床。当数控机床在计算机监控下工作时，它就被称为计算机数控机床（CNC）。

计算机是 CNC 的控制单元，它们内嵌于数控机床中或通过通信通道与数控机床相连。

CNC 就是通过一系列的刀具与工件的动作指令来实现机床或其他加工设备的操作。这些指令的有序组织构成了一个数控程序，该程序可重复使用而获得完全一致的结果。当工作变化了，只需修改程序指令就可得到新的零件，这种能力给 CNC 带来了灵活性。

2.CNC 机床

从最简单到最复杂的机床都会用到数控技术，包括单轴钻床、车床和车削中心、立式或卧式加工中心和铣床、冲床、电火花加工机床、线切割机床、磨床以及测试检测装置等。

3.CNC 的三个组成部分

一般来说，一台计算机 CNC 包括了下面几部分：

1）数控系统

像所有的计算机一样，数控系统只能工作在仅由 0 和 1 组成的二进制下。由一系列 0 和 1 构成的机器语言是计算机唯一能识别的语言。编程的时候，操作人员用一段段的代码和指令来表达工件的信息。

将程序输入后，数控系统通过内部的软件首先将其编译成机器语言，然后计算机将计算各坐标点，并送出控制命令（电子脉冲信号）到伺服驱动系统。

2）伺服驱动系统

伺服驱动系统包括驱动器和伺服电动机，它是连接数控系统和机床本体的桥梁。驱动器将数控装置送出的脉冲信号转换成电动机信号，此信号控制伺服电动机跟随脉冲信号而旋转。伺服电动机旋转反过来又带动丝杠旋转，从而控制各线性轴的运行。随着每一步的运行，检测装置将各轴当前位置信号送回数控装置，数控装置将反馈信号和编程的位置值进行比较从而调整它的输出信号。

3）机床本体

数控机床的机械结构包括两部分。除了主运动部件、进给运动部件（工作台、刀架）、辅助部分（液压、冷却和润滑部分等）和支撑部件（床身、立柱）等一般部件外，还有些特殊部件，如刀库和自动换刀装置等。

通常加工一个零件需要使用几把不同的刀具。加工过程中必须为下一步加工工序迅速换刀。为此，大部分 CNC 机床配备了自动换刀装置。它们无须操作者干预，即可换刀。较为典型的是，自动换刀装置会卡紧车床主轴内的刀具，将其拉出，然后用另一把刀具代替。在具有自动换刀装置的大部分机床上，刀架可以进行旋转，正向或反向均可。

Self-Test

Analyze the following sentences.

(1) For turning operations, the lathe uses a single-point cutting tool which removes metal as it travels past the revolving workpiece.

(2) Precision grinding is concerned with grinding to close tolerances and very smooth finish.

New Words and Phrases

drill *n.* 钻头

vt. 钻孔

lathe *n.* 车床

shaper *n.* 牛头刨床

planer *n.* 龙门刨床

mill *n.* 铣床

grinder *n.* 磨床

twist *n.* 螺旋

ream *v.* 铰孔

tap *v.* 攻丝

screw *n.* 螺钉

bolt *n.* 螺栓

cylindrical *adj.* 圆柱体的，柱面的

axis *n.* 轴

gear *n.* 齿轮

blank *n.* 毛坯；坯料

pulley *n.* 滑轮

shaft *n.* 螺杆

bore *v.* 镗削

groove *n.* 槽口

profile *n.* 侧面

reciprocate *vi.* 往复

tolerance *n.* 公差

Unit 13

Elevator

1. Elevator overview

Elevators are a standard part of any tall commercial or residential building. In recent years, the introduction of the Federal Americans with Disabilities Act has required that many two-story and three-story buildings be retrofitted with elevators.

An elevator, lift in British English, is a type of vertical transport equipment that efficiently moves people or goods between floors of a building. Elevators are generally powered by electric motors or pump hydraulic fluid. In the application of vertical transportation systems, a major decision is which drive system to use, hydraulic or traction? Each type has characteristics which makes it particularly well suited for a specific application. In general, hydraulic elevators are suitable for low-rise buildings (up to 6 floors) whereas, the roped elevators (or "traction elevator" as Figure 13.1 shows) are best suited to higher buildings.

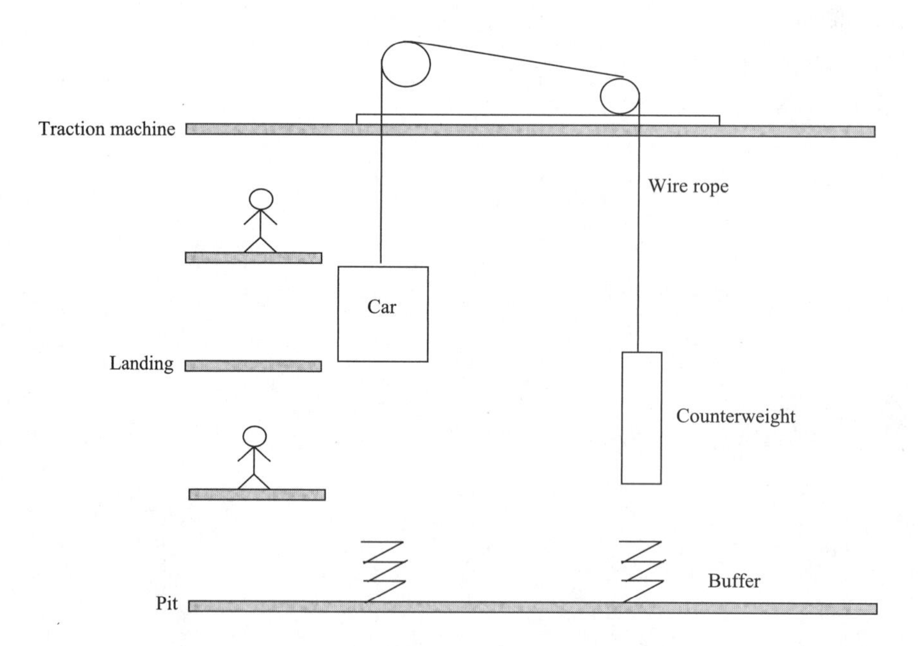

Figure 13.1 Traction elevator

Traction elevators are the most popular type nowadays and are driven by the traction between the suspension ropes and drive sheaves.

Elevators themselves are simple devices, and the basic lifting systems have not changed much in the 50 years. In space, elevators can be considered to be composed of four pasts: machine room, landing, shaft and car (as shown in Figure13.2).

(a) Machine room

(b) Landing

(c) Shaft

(d) Car

Figure 13.2 Four parts of space for elevator

The machine-room-less elevator (as Figure 13.3 shows) is the result of technological advancements. These newly designed permanent magnet motors (PMM) allow the manufacturers to locate the machines in the hoistway overhead thus eliminating the need for a machine room over the hoistway. This design has been utilized for at least 15 years and is becoming the standard product for low to mid rise buildings. It was first introduced to the U.S. market by KONE.

According to their function, elevator is made up of eight system: traction system, guide system, car system, door system, weight balance system, electrical drive system, electrical control system and safety protection system (as Figure 13.4 shows)

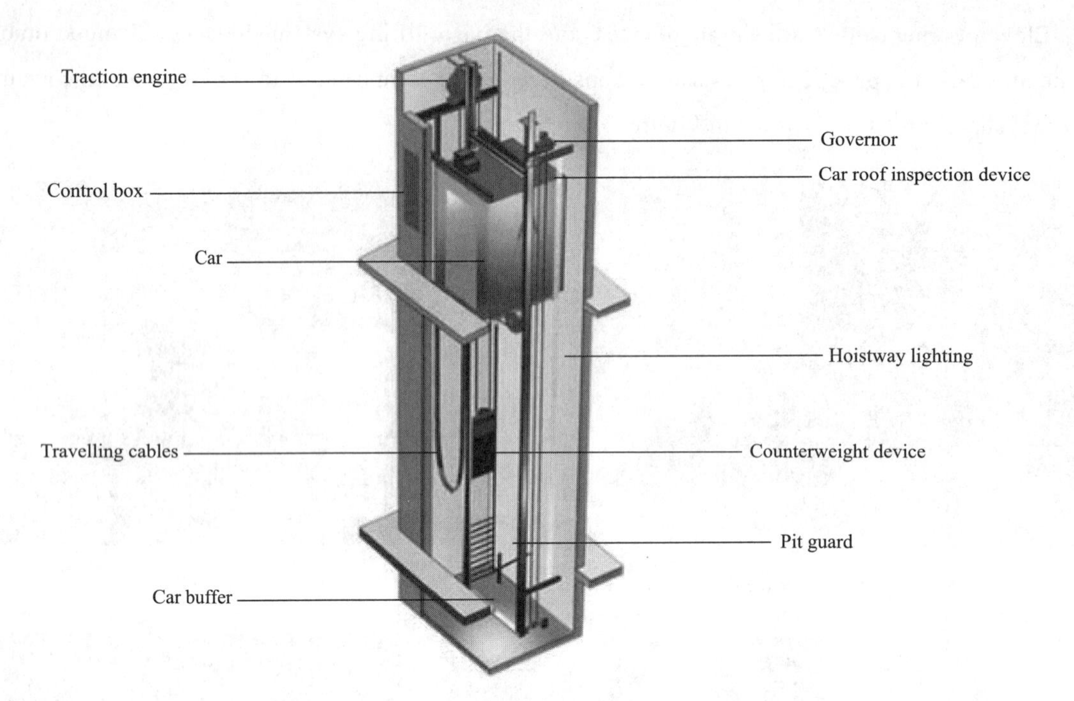

Figure 13.3 Machine-room-less elevator

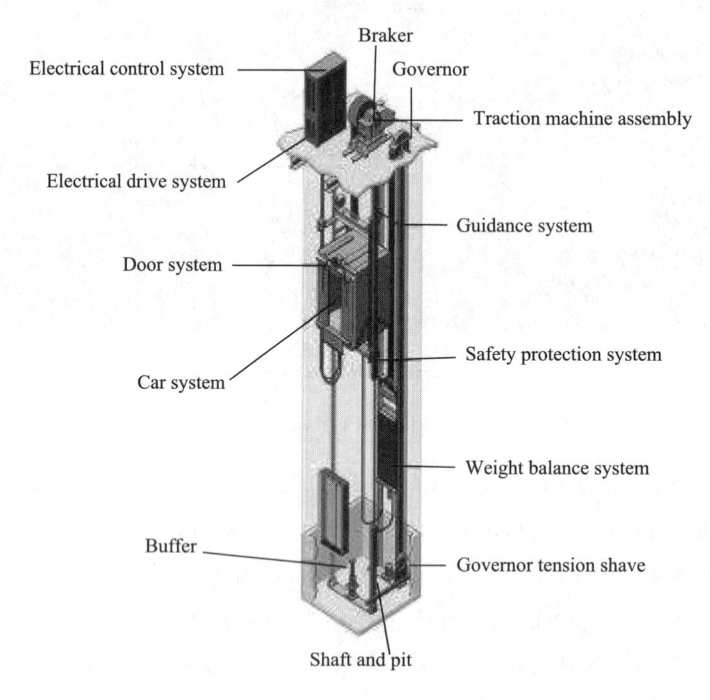

Figure 13.4 Traction elevator overview

2. Traction system

The most popular elevator design is the traction elevator (or "roped elevator"). Traction elevators are much more versatile than hydraulic elevators, as well as more efficient. In traction elevators (as Figure 13.5 shows), the car is raised and lowered by traction steel ropes rather than pushed from below. The ropes are attached to the elevator car, and looped around the traction sheave. A traction sheave is just a pulley with grooves around the circumference. The sheave grips the hoist ropes, so when you rotate the sheave, the ropes moves too.

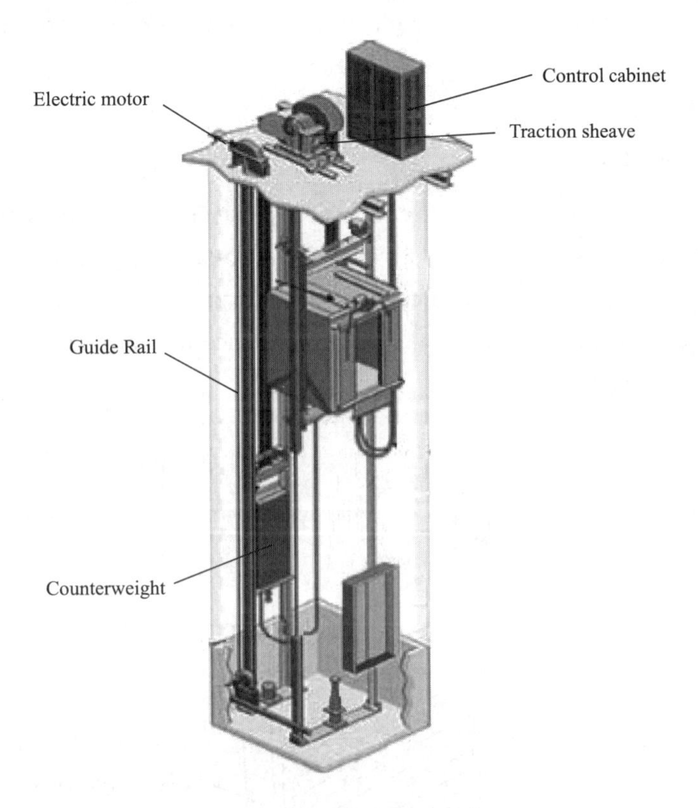

Figure 13.5 Traction system

The sheave is connected to an electric motor. When the motor turns one way, the sheave raises the elevator; when the motor turns the other way, the sheave lowers the elevator. In gearless elevators, the motor rotates the sheaves directly. In geared elevators, the motor turns a gear train that rotates the sheave. Typically, the sheave, the motor and the control cabinet are all housed in a machine room above the elevator shaft.

The ropes that lift the car are also connected to a counterweight, which hangs on the other side of the sheave. The counterweight weighs about the same as the car filled to 40-percent capacity. In other words, when the car is 40 percent full, the counterweight and the car are perfectly balanced.

The purpose of this balance is to conserve energy. With equal loads on each side of the sheave, it only takes a little bit of force to tip the balance one way or the other. Basically, the motor only has to overcome friction—the weight on the other side does most of the work. To put it another way, the balance maintains a near constant potential energy level in the system as a whole. Using up the potential energy in the elevator car (letting it descend to the ground) builds up the potential energy in the weight (the weight rises to the top of the shaft). The same thing happens in reverse when the elevator goes up. The system is just like a see-saw that has an equally heavy kid on each end.

Both the elevator car and the counterweight ride on guide rail along the sides of the elevator shaft. The rail keeps the car and counterweight from swaying back and forth, and they also work with the safety system to stop the car in an emergency.

3. Weight balance system

1) Counterweight

The counterweight (as show in Figure 13.6) on most elevators hangs from its own cable attached to the elevator. That cable travels from the car, over pulleys at the top of the elevator shaft, and down to the counterweight. The counterweight is usually equal to the mass of the empty elevator is 40% filled, the counterweight will exactly balance the car and very little work will be done in raising or lowering the car.

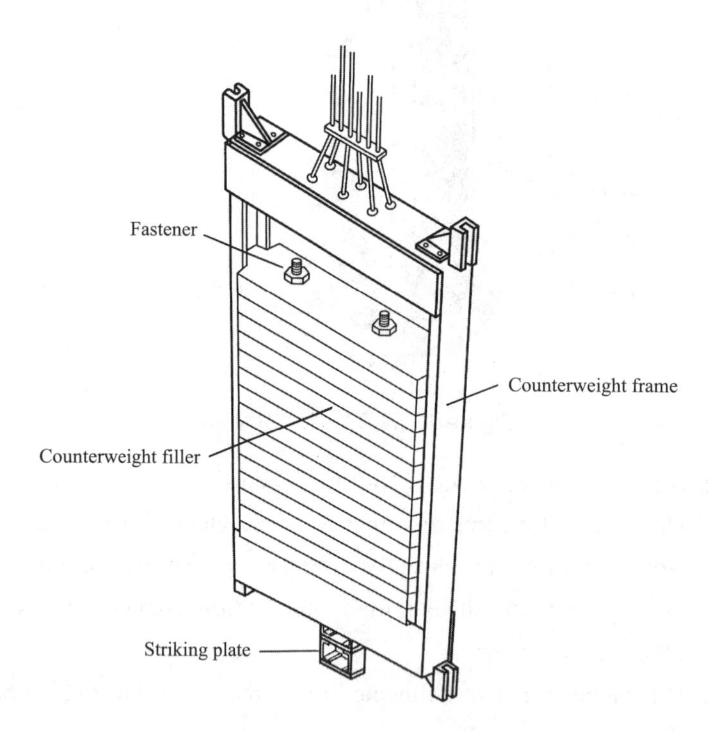

Figure 13.6 Counterweight

2) Weight compensation device

Elevators with more than 100 ft (30 m) of travel should have a weight compensation system (see Figure 13.7). This is separate set of cables or a chain attached to the bottom of the counterweight and the bottom of the elevator cab. This makes it easier to control the elevator, as it compensates for the differing weight of cable between the counterweight side and the car side cause by the steel wire rope.

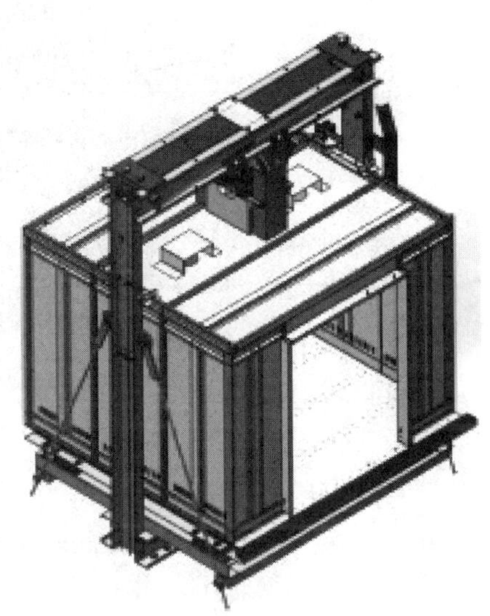

Figure 13.7 Weight compensation system

4. Car and door system

1) Car

Car is raised and lowered mechanically in a vertical shaft in order to move people or goods floor to another in a build. In a "traction" lift, cars are pulled up by means of rolling steel ropes over a deeply grooved pulley, commonly called a sheave in the industry. The weight of the car balanced by a counterweight.

Car can be divided into two major parts:car body and car frame. Car body (as shown in Figure 13.8) is composed of car floor, car walls and car roof. The supporting frame (as shown in Figure 13.9) of an elevator car to which are attached the car platform, guide shoes, elevator car safety, hoisting ropes or sheaves, and/or associated equipment.

The crosshead is the upper member of the car frame. The stiles are the vertical member of the car sling, one on each side, which fastens the crosshead to the safety plank.

The brace rod is a rod extending from the elevator platform framing to another part of the elevator car frame or sling for the purpose of supporting the platform or holding it securely in position. Brace rods are support for the outer corners of the platform, each of which tie to upper portions of the stile.

The platform isolation is rubber or other vibration absorbing material which reduces the transmission of vibration and noise to the platform. These pads are often replaced when modernizing as new isolation is more resilient and helps to reduce vibration and improve the comfort of the ride for passengers.

The bolster is the bottom horizontal member of a hydraulic car sling, to which the platen plate attaches. The safety plank for a traction elevator, bottom member of a sling, contains the safety.

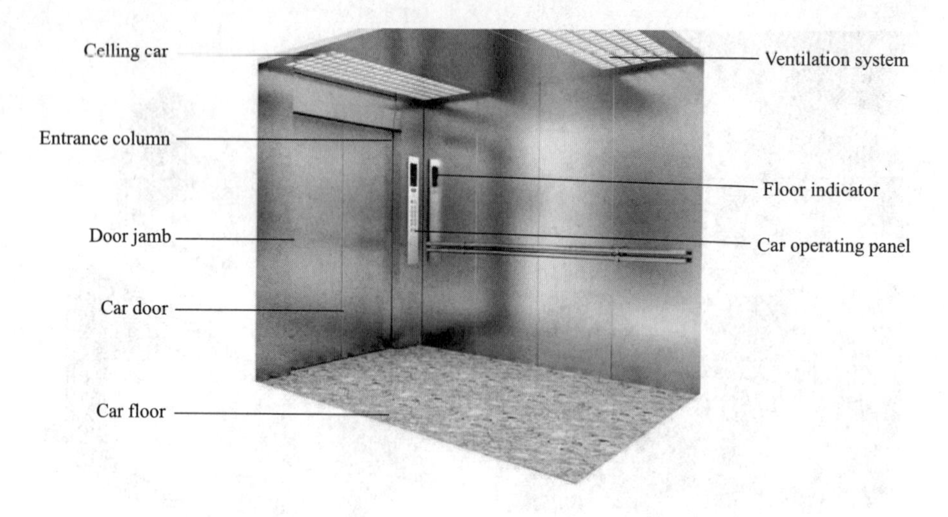

Figure 13.8 Car body

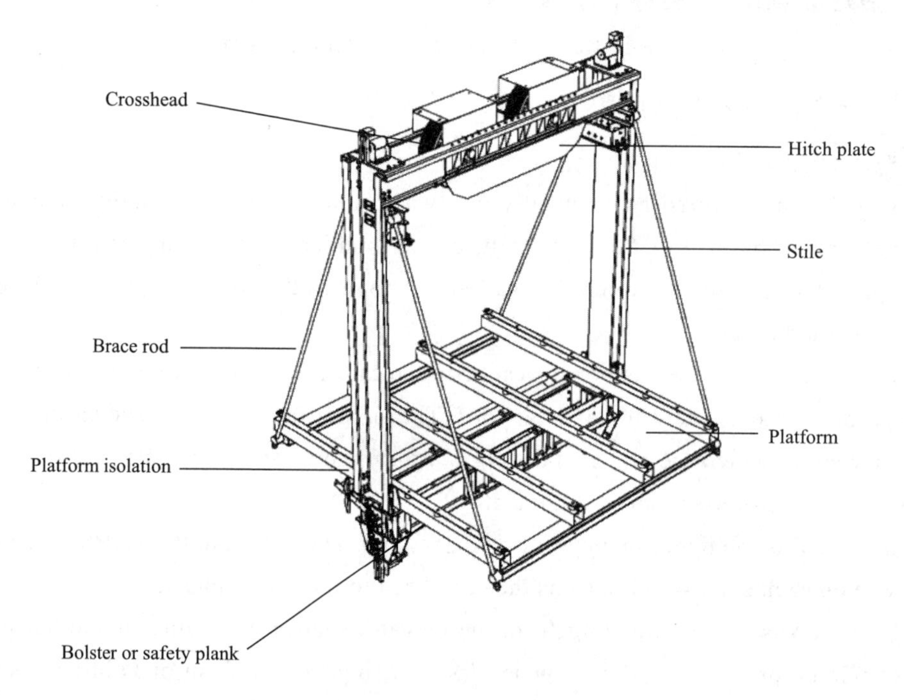

Figure 13.9 Car frame

2) Car door and landing door

When we talk about elevator door, there are car door and landing door which are two different conceptions. Before you enter elevator, you face landing door; when you stay in elevator car, the door you can see is car door. Each elevator has one pair of car door but many sets landing door depending on the height of building. Car door is active door while landing door is passive door.

Landing door protect riders from falling into the shaft. Figure 13.10 depicts three major types of landing doors used in elevator. The passenger types are the center-opening, the single slide and the two-speed sliding door (two panels overlapped when open). The typical car landing door is the vertical bi-parting, the top half rises and the lower half depresses at the same time to form an opening almost as wide as the car itself.

(a) Vertical biparting

(b) Single-slide

(c) Center-opening

Figure 13.10 Landing doors

5. Guide system

1) Guide rail and rail bracket

Elevator guide rail used to guarantee the car or counterweight move up and down without much swing is supported by rail bracket (as Figure 13.11 shows) mounted on the wall of shaft. The actual most used type is T shape rail (as Figure 13.12 shows) and hollow rail (as Figure 13.13 shows) can be used in counterweight side or freight elevator. Guide rail will suffer huge load when safety gear operation, so its rigidity or stiffness is important.

Figure 13.11 Rail bracket

Figure 13.12 T shape rail

Figure 13.13 Hollow rail

2) Guide shoe

The guide shoes used to guide the car and counterweight on the guide rail may be either roller or sliding type, as shown in Figure 13.4 and Figure 13.5. Sliding guide shoes are often used for rated speed up to 150 ft/min (about 0.76 m/s). In some cases, heavy duty freight elevator may use sliding guide shoe for higher speed.

Figure 13.14 Roller guide shoe

Figure13.15 Sliding guide shoe

6. Safety protection system

Elevators are used by thousands of people every day. Responsible building owners all over the world are doing their part to keep people safe by upgrading their existing vertical transportation equipment to meet all local and national safety codes.

1) Overspeed governor (as shown in Figure 13.6)

The governor safety system is an independent safety system that is not intended to regulate speed or stop the elevator during normal operation. The movement of the car passes to the governor and will not operate it unless the car overspeed.

Figure 13.16 Overspeed governor

2) Safety gear

Safety gear (see Figure 13.17) is braking system attached on the elevator car, running up and down the elevator shaft and grabs onto the rails when emergency. Some safeties clamp the rails, while others drive a wedge into notches in the rails. Typically, safeties are activated by a mechanical speed governor.

Figure 13.17 Safety gear

3) Rope gripper

The primary overspeed governor and safety can protect the car from falling to the pit, but it will not stop the ascending car. In order to stop the ascending car, we now have another device, called rope gripper, as shown in Figure 13.18.

4) Buffer

Buffer (see Figure 13.19) is an important part of elevator safety system, and it is the last barrier if elevator out of control. Buffers can reduce the shock to a great extent or soften the force when the elevator runs into the pit during an emergency, protect passengers from being harmed or components being destroyed.

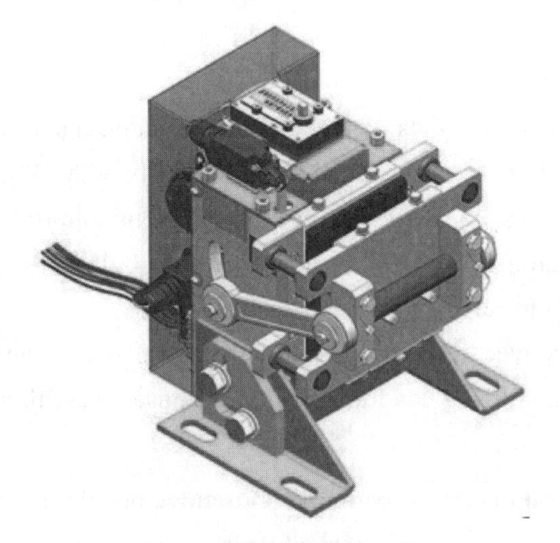

Figure 13.18 Rope gripper

Figure 13.19 Buffer

7. Electrical control system

Automatic elevators began to appear as early as the 1930s, their development being hastened by skyscrapers increase in large cities such as New York and Chicago. These electromechanical systems used relay circuits to control the speed, position and door operation of an elevator. Relay-controlled elevator systems remained common until the 1980s, and their gradual replacement with microprocessor based controls which are now the industry standard. Elevator electrical control system includes car operation panel, control cabinet, calling board, etc., as shown in Figure 13.20 to Figure 13.22.

Figure 13.20 Car operation panel

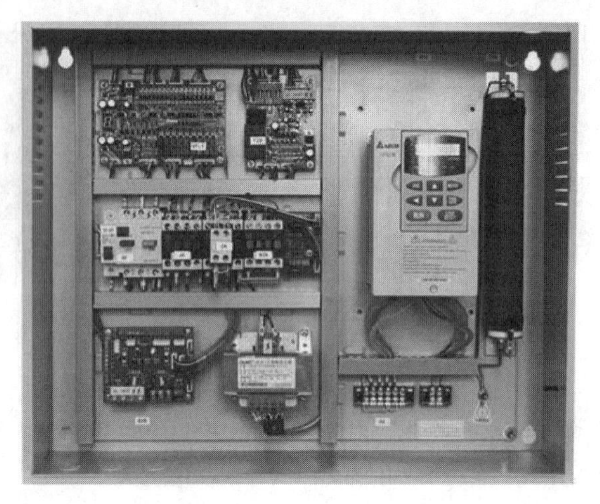

Figure13.21 Control cabinet

Figure13.22 Calling board

Extension Materials

Moving walkways types

Moving walkways has many of the same safety design feature as do escalators and should be treated in the same manner. Moving walkways have a maximum slope of 12 degrees (escalators are usually 30 degrees plus or minus 1 degree). There are two types of moving walkway being manufactures today：pallet and continuous belt.

Pallet types escalator consists of many separate pallets connected with a chain, much like an escalator. The pallet may be die-cast aluminum or stainless sheet-metal, widths are between 32 inches (800 mm) and 56 inches (1,200 mm). They are usually with a speed of feet per minute (0.5 m/s), powered by an AC induction motor. Continuous belt type: A heavy rubber belt is used and supported on each side with rollers below the belt, referred to as edge support.

Both types of moving walkways have a grooved surface to mesh with combplates at the ends. Also, nearly all moving walkways are built with moving handrails similar to those on escalators.

1. Inclined moving walkways

An inclined moving walkway is used in airports and supermarkets to move people to another floor with the convenience of an elevator (namely that people can take along their suitcase trolley or shopping cart, or baby carriage) and the capacity of an escalator.

The carts have either a brake that is automatically applied when the cart handle is released, strong magnets in the wheels to stay adhered to the floor, or specially designed wheels that secure the cart within the grooves of the ramp, so that wheeled items travel alongside the riders and do not slip away.

2. High-speed walkways

In the 1970s, Dunlop developed the speedways system. The great advantage of the speedway, as compared to the then existing systems, was that the embarking/disembarking zone was both wide and slow moving (up to 4 passengers could embark simultaneously, allowing for a large number of passengers, up to 10,000 per hour), whereas the transportation zone was narrower and fast moving.

The entrance to the system was like a very wide escalator, with broad metal tread plates of a parallelogram shape. After a short distance the tread plates were accelerated to one side, sliding past one another to form progressively into a narrower but faster moving track which travelled at almost a right-angle to the entry section. The passenger was accelerated through a parabolic path to maximum design speed of 15 km/h (9 mph).

The experience was unfamiliar to passengers, who needed to understand how to use the system to be able to do so safely. Developing a moving handrail for the system presented a challenge, also solved by the Battelle team. The speedway was intended to be used as a stand alone system over short distances or to form acceleration and deceleration units providing entry and exit means for a parallel conventional walkway. The system was still in development in 1975 but never went into commercial production.

In 2002, the first successful high-speed walk was installed in the Montparnasse station in Paris. At first, it operated at 12 km/h (7 mph) but due to people losing their balance, the speed was reduced to 9 km/h (6 mph). It has been estimated that commuters using a walkway such as this twice a day would save 15 minutes per week and 10 hours a year.

Using the high-speed walkway is like using any other moving walkway, except that for safety there are special procedures to follow when joining or leaving. When this walkway was introduced, staff in yellow jackets determined who could and who could not use it. As riders must have at least one hand free to hold the handrail, those carrying bags, shopping, etc., or who are infirm, must use the ordinary walkway nearby.

自动人行道的类型

自动人行道具有很多与扶梯相同的安全设计，应等同对待。自动人行道斜度最大 12°（扶梯一般是 30° ±1°）。主要有两种类型的自动人行道在生产：托盘式和连续带式。

托盘式扶梯包含多块用链条连接在一起的独立托盘，与扶梯非常类似。托盘可以是铸铝或者不锈钢的，宽度在 32 英寸（800 mm）和 56 英寸（1 200 mm）之间。速度一般为 100 英尺 /min（0.5 m/s），由交流感应电动机驱动。连续带式：用一种厚橡胶传动带在扶梯两侧通过传动带下方的棍子支撑，称为侧边支撑。

两种类型的自动人行道都具有沟槽表面，在扶梯末端同梳齿板相啮合。而且，几乎所有的自动人行道都具有与扶梯类似的移动扶手带。

1. 倾斜式自动人行道

倾斜式自动人行道用于机场或超市运送人员到另一楼层，具有电梯的便利（也就是人们可以随身携带行李车或购物车、童车）和扶梯的输送能力。

手推车或者具备制动装置，制动装置在手推车把手释放时自动启用，牢固的磁力让轮子附着在底板上，或者采用特制的轮子使手推车处于斜坡槽中，以便轮子骑在凸起处不至于滑走。

2. 高速自动人行道

在 20 世纪 70 年代，邓洛开发了该高速系统。相比于现存的系统，高速系统的最大优势在于出入区域宽而且移动慢（四名乘客可同时进入，最大允许乘客量达每小时 10 000 人），但是运送区域较窄而且移动速度快。

系统入口类似一个非常宽的扶梯，具有平行四边形形状的宽大金属踏板。在短距离后，踏板被加速到一边，相互滑动，逐渐形成一个更窄但更快的移动轨道，几乎以直角行驶到入口部分。乘客经过一个抛物线路径被加速到自动人行道的最大设计速度 15 km/h (9 mph)。

这种体验对乘客来说并不熟悉，他们需要理解如何使用该系统是安全的。为现在的系统开发一种移动扶手带是一种挑战，但已被巴泰尔团队所解决。该高速通道打算用于短途独立系统，为传统与之并行的人行道出入口提供加速和减速的方法。该系统开发于 1975 年，但从未投入商业生产。

2002 年，第一条成功的高速通道安装在巴黎的蒙帕纳斯地铁站。起初运转速度为 12 km/h (7 mph)，由于人们易失去平衡，速度降低到 9 km/h (6 mph)。据估计，经常乘车的人如果每天使用该通道两次，每周会节省 15 min，每年会节省 10 h 。

使用高速通道跟其他自动人行道相同，除了为安全而设的进入及离开时的特殊步骤。刚引进该通道时，穿黄色夹克的工作人员决定谁能和谁不能使用该通道。行人必须至少一只手空闲，以便握住扶手带，而那些携带箱包、购物之类的人，或体弱的人必须使用旁边的普通通道。

Self-Test

Complete the following topics according to what you have learned.

(1) In general, hydraulic elevators are suitable for low-rise buildings whereas, the traction elevators are best suited to higher buildings ( ).

A . True　　B. False　　C. Maybe　　D. Not always

(2) The brake on the driving machine ( ).

A. serves to stop the elevator car

B. only sets when the mainline switch is opened

C. is applied by spring force when electric power is removed

D. is attached to the crosshead

(3) The purpose of the overspeed governor is to ( ).

A. control the speed of the elevator

B. cause the power to removed and the safety to set when overspeed occurs

C. assure that the elevator is operating as fast as possible

D. signal the controller to cause the elevator to move

(4) Standing inside the elevator, what can you see? ( ).

A. Landing door

B. Car door

C. Both car door and landing door

D. Neither car door nor landing door

New Words and Phrases

commercial or residential building　商业或住宅建筑

retrofit　*v.*　改进；更新；改装

　　n.　式样翻新

PMM (permanent magnet motor)　永磁电动机

MRL (machine-room-less) elevator　无机房乘客电梯

traction　*n.*　曳引；牵引

resilient　*adj.*　有弹力的

stile　*n.*　立梁，侧板；立柱

bi-parting　*adj.*　上下对开的；垂直中分的

category　*n.*　种类，分类，范畴

commuter　*n.*　往返于两地之间的人

walkway　*n.*　人行道，通道，走廊

infirm　*adj.*　体弱的

parallel　*n.*　平行线；对比

　　adj.　平行的；类似的

Unit 14

Robot

1. Introduction of robot

Robots are used everywhere not only in working places but also at home. These machines can do work that is no dangerous for humans, help around the house, or can be used for fun.

1) Laws of robotics

Laws of robotics are a set of laws, rules, or principles, which are intended as a fundamental framework to underpin the behavior of robots designed to have a degree of autonomy. The best known set of laws is Isaac Asimov's "Three Laws of Robotics" (often shortened to The Three Laws or Asimov's Laws) in 1942. The Three Laws are described as below:

First: A robot may not injure a human being or, through inaction, allow a human being to come to harm.

Second: A robot must obey the orders given it by human beings except where such orders would conflict with the First Law.

Third: A robot must protect its own existence as long as such protection does not conflict with the First or Second Laws

In later fiction, Asimov added a Zeroth Law. To precede the others:

A robot may not injure humanity, or, by inaction, allow humanity to come to harm.

2) Types and application of robot

Robotics is a rapidly growing field with the continuous advancement of technology. Now, robots can be found in homes as toys, vacuums, and programmable pets. Also, robots are an important part of many aspects of industry, medicine, science, space exploration, construction, food packaging and are even used to perform surgery.

There are many types of robots, such as mobile robots, industrial robots (manipulating), service robots, educational robots, modular robots, collaborative robots, etc., as shown in Figure 14.1. They are used in many different environments and for many different purposes. Robots have replaced humans in performing repetitive and dangerous tasks, which humans prefer not to do, or are unable to do because of size limitation, or which take place in extreme environments such as outer space or the bottom of the sea.

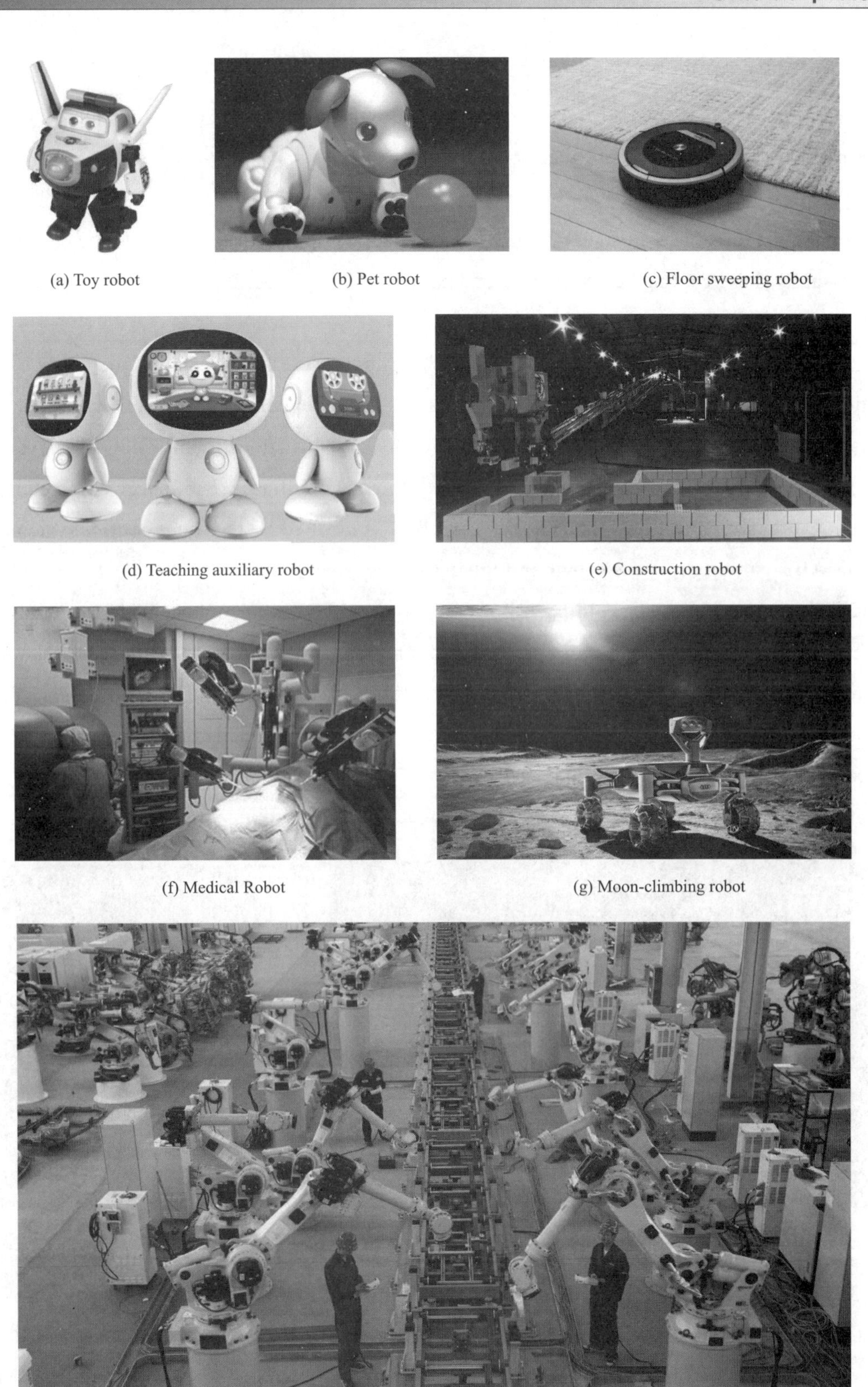

(a) Toy robot (b) Pet robot (c) Floor sweeping robot

(d) Teaching auxiliary robot (e) Construction robot

(f) Medical Robot (g) Moon-climbing robot

(h) Industrial robot

Figure 14.1 Robots for various applications

3) Industry application of robot

(1) Painting robot. "Painting robot" is an industry term for a robot that has two major differences from all other standard industrial robots.

① Explosion proof arms. Painting robots are built with explosion proof robot arms, meaning that they are manufactured in such way that they can safely spray coatings that create combustible gases. Usually these coatings are solvent based paints, when applied, it must create an environment that must be monitored for fire safety.

② Self-contained paint systems. When painting robots were first designed, they only had one function—to work safely in a volatile environment. As acceptance and use expanded, painting robots grew into a unique subset of industrial robots, not just traditional robot with explosion proof options. Painting robots now have the ability to control all aspects of spray parameters. Fan air, atomization air, fluid flow, voltage, etc., can all be controlled by the robot control system.

The ABB painting robot is shown in Figure 14.2.

(2) Welding robot. The welding robot (see Figure 14.3) is an industrial robot engaged in welding. Industrial robot is a versatile, reproducible manipulator with three or more programmable axes for industrial automation. In order to adapt itself to different uses, the mechanical interface of the robot's last axis is usually a connecting flange, which can be attached to different tools or end effectors. The welding robot is an industrial robot of which a welding tongs or torch is in the end, so that it can weld, cut or hot spray.

Figure 14.2 The ABB painting robot

Figure 14.3 Welding robot

(3) Handling robot.Handling robots (see Figure 14.4 and Figure 14.5) are industrial robots can be automated for handling operation. There are tandem joint robots, horizontal joint robots (SCARA robots), Delta parallel joint robots and AGV handling robots in common. Above all kinds of robots in the handling industry have their own characteristics.

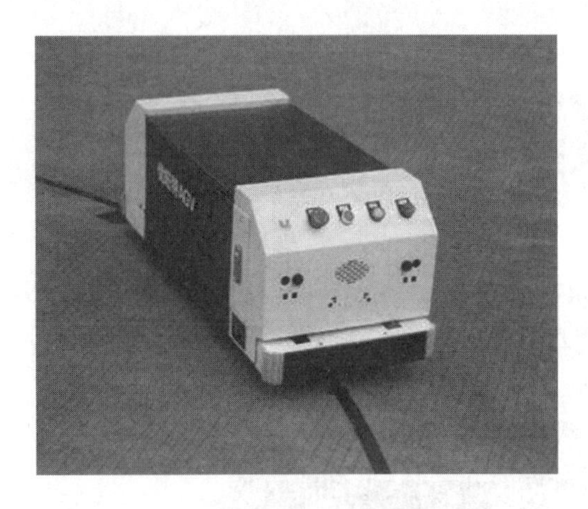

Figure 14.4 AGV

Figure 14.5 Palletizing handling robot

(4) Assembly robot (see Figure 14.6). "Assembly robot" is an industry term for a robot that is used in lines of industrial automation production to assemble parts or components and it is the core of the flexible assembly system automation equipment. It's used for lean industrial processes and has expanded production capabilities in the manufacturing world. The assembly line robot can increase production speed and consistency. The tools at the end of arm can be customized for each assembly robot to cater to the manufacturing requirements and used for all applications.

Figure 14.6 Assembly robot

(5) Polishing robot (see Figure 14.7). "Polishing robot" is an industry term for robots that can polish automatically. It is widely used in the fields of 3C, sanitary ware, IT, auto parts, industrial parts, medical devices and civil products.

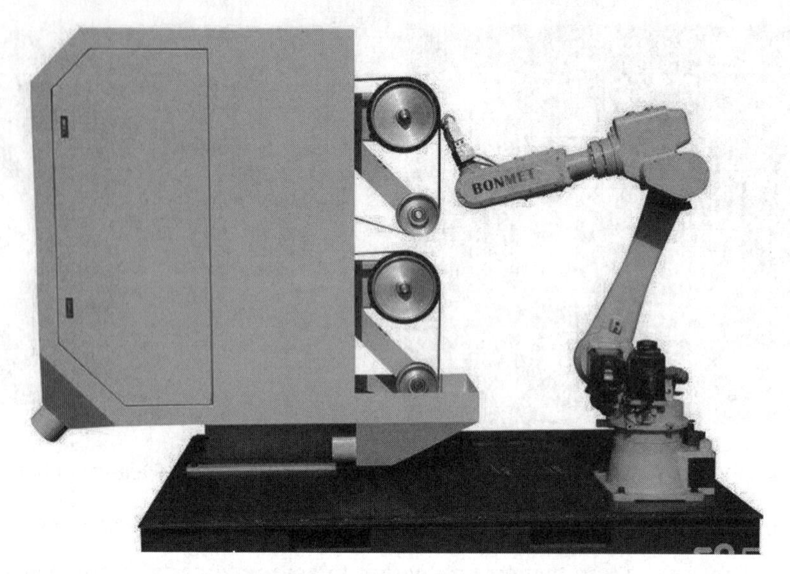

Figure 14.7 Polishing robot

2. About industrial robot

1) Definition and application

Industrial robot is a robot system used for manufacturing. Industrial robot is defined as "an automatically controlled, reprogrammable, multipurpose, manipulator programmable in three or more axes, which may be either fixed in place or mobile for use in industrial automation applications." in ISO 8373.

The Robotics Institute of America (RIA) defines a robot as: A reprogrammable multifunctional manipulator designed to move materials, parts, tools, or specialized devices through variable programmed motions for the performance of a variety of tasks.

Industrial robot in helpful in material handling and providing interface. Typical applications of robots include welding, painting, assembly pick and place for PCB, packaging, labeling, palletizing, inspection and testing. All can be accomplished with high endurance, speed, and precision.

2) Main parts of industrial robot

Robots can be made from a variety of materials including metals and plastics. The industrial robot is composed of four parts: manipulator, controller, demonstrator and robotic hand, as shown in Figure 14.8.

(1) Manipulator is also known as the robotic arm, a type of mechanical arm, usually programmable, with similar functions to a human arm, The arm may be the sum total of the mechanism or may be part of a more complex robot. Robotic arm is connected by joints, which allow either rotational motion or translational (linear) displacement. The terminus of the manipulator is an end effector.

(2) Controller is also known as the "brain" which is run by a computer program. Usually the program is very detailed as it gives commands for the moving parts of the robot to follow.

(3) Demonstrator is also known as the teaching box, a human-computer interaction interface, connected with the controller, can be operated to move by the operator.

(4) Robotic hand is the end effector, or end-of arm tooling (EOAT), can be designed to perform any desired tasks such as welding, gripping, spinning, etc., depending on the application. End effectors are generally highly complex, made to match the handled products and often capable of picking up an array of products at one time. They may utilize various sensors to aid the robe system in locating, handling, and positioning products.

All of these parts work together to control how the robot operates.

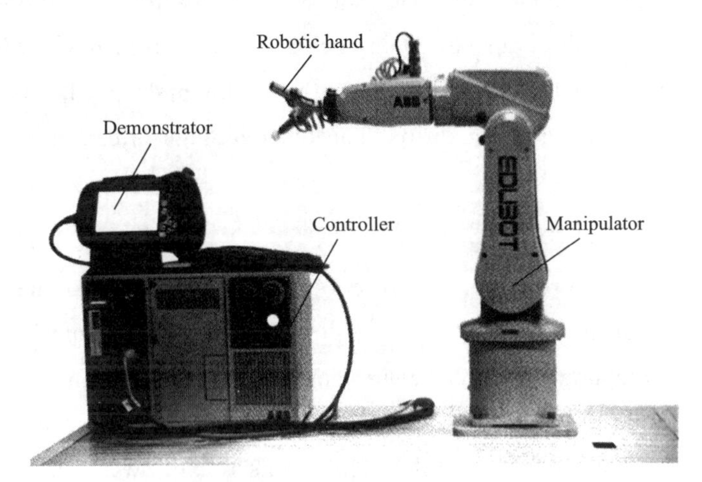

Figure 14.8 Main parts of industrial robot

3) Basic terms

(1) Rigid body. In physics, a rigid body is an idealization of a solid body whose deformation can be neglected. In other words, the distance between any two given points of a rigid body remains constant all the time regardless of external forces exerted on it.

(2) Rotary joint. A rotary joint (also called pin joint or hinge joint) is a one degree of freedom kinematic pair used in mechanisms. Rotary joints provide single axis rotation function used in many places such as door hinges, folding mechanisms, and other uni-axial rotation devices.

(3) Kinematic pair. A kinematic pair is a connection between two bodies that impose constraints on their relative movements.

(4) Articulated robot. Articulated robot is the one that uses rotary joints to access. its work space. Usually the joints are arranged in a "chain", so that one joint supports another further in the chain path.

(5) Continuous robot. It is a control scheme by the inputs or commands that specifies every point moves long a desired path of motion. The path is controlled by the coordinated motion of the manipulator joints.

(6) Kinematics. The actual arrangement of rigid members and joints in the robot determines the robots possible motions. The kinds of robot kinematics include articulated, Cartesian, parallel and SCARA.

(7) Motion control. For some applications, such as simple pick and place, assembly, the robot need merely return repeatedly to a limited number of pre-taught positions, For more sophisticated applications, such as welding and finishing (spray painting), motion must be continuously controlled to follow a path in space controlled orientation and velocity.

(8) Power source. Some robots use electric motors, others use hydraulic actuators. The former are faster, the latter are stronger and more advantageous in applications such as spray painting, when a spark could set off an explosion.

(9) Drive. Some robots connect electric motors to the joints via gears, others connect the motor to the joints directly (direct drive). Using gears results in measurable "backlash" which is free movement in an axis. Small robot arms frequently employ high speed, low torque DC motors, which generally require high gearing ratios. The disadvantage is backlash. In such cases the harmonic drive is often used.

4) Types of industrial robots

(1) Cartesian coordinate robot (see Figure 14.9). Cartesian coordinate robot (also called linear robot) is an industrial robot with simple structure. A popular application for Cartesian coordinate robot is a computer numerical control machine (CNC machine) and 3D printing (see Figure 14.10). The simplest application is used in milling and drawing machines where a pen or a router translates across an *X-Y* plane while a tool is raised and lowered onto a surface to create a precise design. Pick and place machines (see Figure 14.11) and plotters are also based on the principle of Cartesian coordinate robot.

(2) SCARA robot. (see Figure 14.12) The SCARA acronym stands for selective compliance assembly robot arm or selective compliance articulated robot arm.

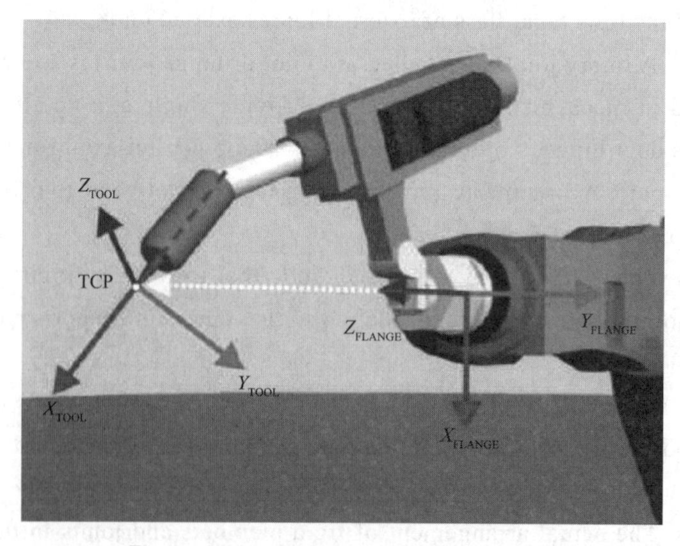

Figure 14.9 Cartesian coordinate robot

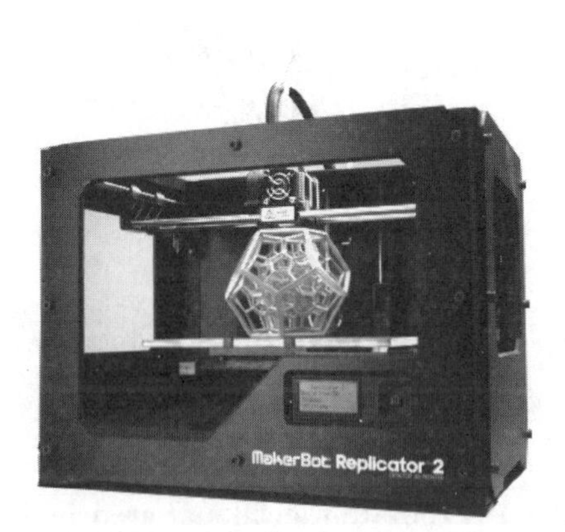

Figure 14.10 3D printing

Figure 14.11 Pick and place machine

(3) Six-axis articulated robot (see Figure 14.13). The six-axis articulated robot is a kind of widely used mechanical equipment. The vast majority of articulated robots have six axes, also called six degrees of freedom. Six-axis robots allow for greater flexibility and can perform a wider variety of application than robots with fewer axes.

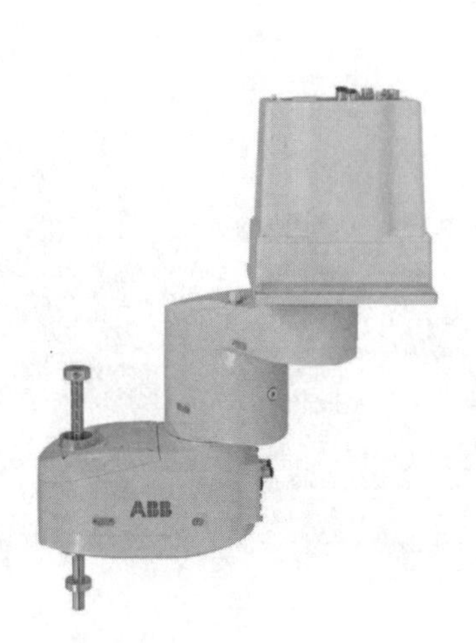

Figure 14.12 SCARA robot

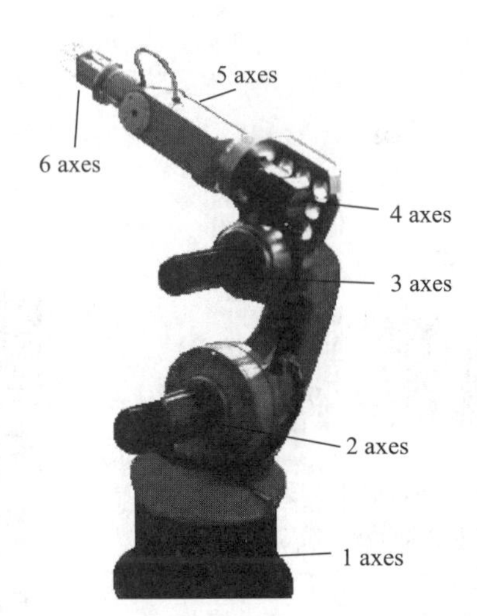

Figure 14.13 Six-axis articulated robot

(4) Palletizing robot (see Figure 14.14). Industrial palletizing refers to loading and unloading parts, boxes or other items to or from pallets.

(5) Delta robot (see Figure 14.15). The Delta robot is a type of parallel robot. It consists of three arms connected to universal joints at the base. The robot can also be seen as a spatial generalization of a four-bar linkage.

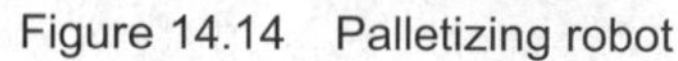

Figure 14.14 Palletizing robot

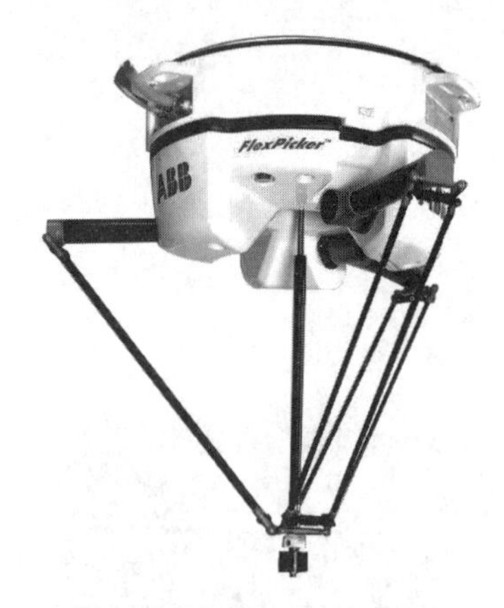

Figure 14.15 Delta robot

3. The companies that make robots

1) ABB and ABB Robot

ABB (ASEA Brown Boveri) is a Swiss multinational corporation headquartered in Zurich, Switzerland, operating mainly in robotics and the power and automation technology areas. ABB resulted from the merger of the Swedish corporation Allma-nna Svenska Elektriska Aktiebolaget(ASEA) and the Swiss company Brown, Boveri & Cie(BBC) in 1988, as shown in Figure 14.16. ABB is one of the largest engineering companies, its core businesses is power and automation technologies.

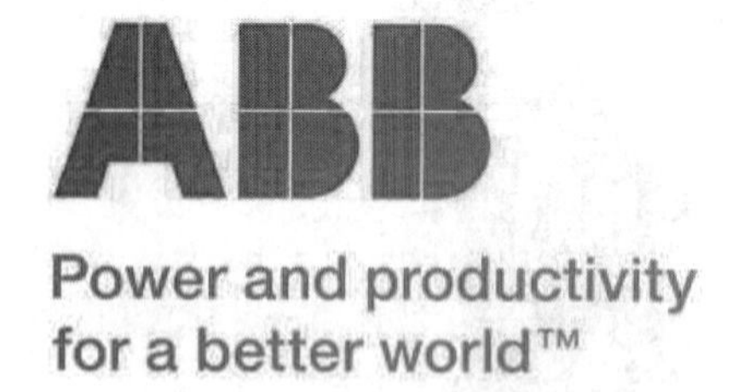

Figure 14.16 ABB

As shown in Figure 14.17, ABB expands RobotStudio® with ability to control SCARA robots from PC. RobotStudio's new Robot Control Mate add-on allows users to easily commission and control SCARA robots direct from their laptop and computers ABB has simplified the use of its popular SCARA robots by giving users the ability to commission and control the robots from their PC through RobotStudio®, ABB's simulation and offline programming software.

Robot Control Mate is an add-on to RobotStudio, enabling users to jog, teach and calibrate robots from their computer, making it easier than ever to control a SCARA robot's movements.

For the first time, ABB's offline programming software can be used to control the physical movements of a robot in real-time. Robot Control Mate also makes robot programming possible when a FlexPendant is not in place.

Robot Control Mate will first be available on the IRB 910INV ceiling-mounted robot and will be expanded to other robots in ABB's portfolio later years.

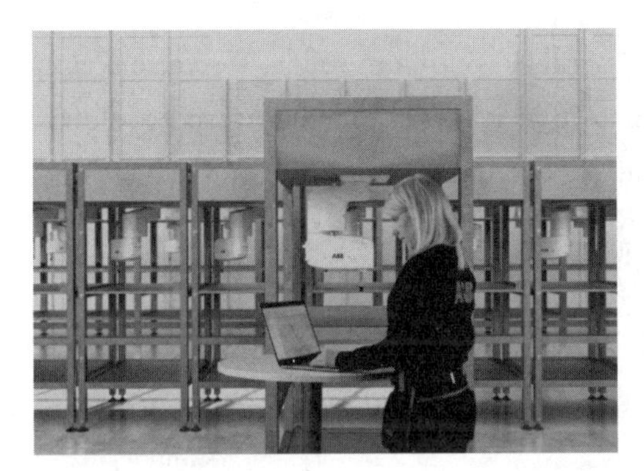

Figure 14.17 ABB robot

2) KUKA and KUKA robot

KUKA is a German manufacturer of industrial robots and solutions for factory automation. KUKA is one of the world's leading suppliers of automation solutions.

The corporate history of KUKA began in 1898 with Johann Joseph Keller and Jakob Knappich in Augsburg and KUKA is created from the first letters of "Keller und Knappich Augsburg". For more than 100 years. KUKA has stood for ideas and innovations that have made it successful worldwide. More than 12,000 KUKA colleagues worldwide are developing the intelligent robot-based automation solutions of tomorrow creative integrating and effective. It is called "Orange Intelligent". At the interface between the virtual and real worlds, "rang intelligent" is the creative, integrating and effective power of KUKA.

Today, KUKA stands for innovations in automation and is a driver of Industry 4.0. KUKA robots is shown in Figure 14.18.

Figure 14.18 KUKA

3) YASKAWA and YASKAWA robot

The YASKAWA Electric Corporation is the world's largest manufacturer of AC inverter drives, servo and motion control, and robotic automation systems company was founded in 1915, and its head office is located in Japan.

YASKAWA has successively commercialized and marketed optimum robots for various uses since 1977, centering on arc welding, one of its areas of expertise, and including spot welding, handling, assembly, painting, transfer of liquid crystal panels, transfer of semiconductor wafers and so on.

Figure 14.19 shows Different series of YASKAWA robots. For example, YASKAWA MH12 robot is generally composed of three parts: manipulator, DX200 controller and programming pendant.

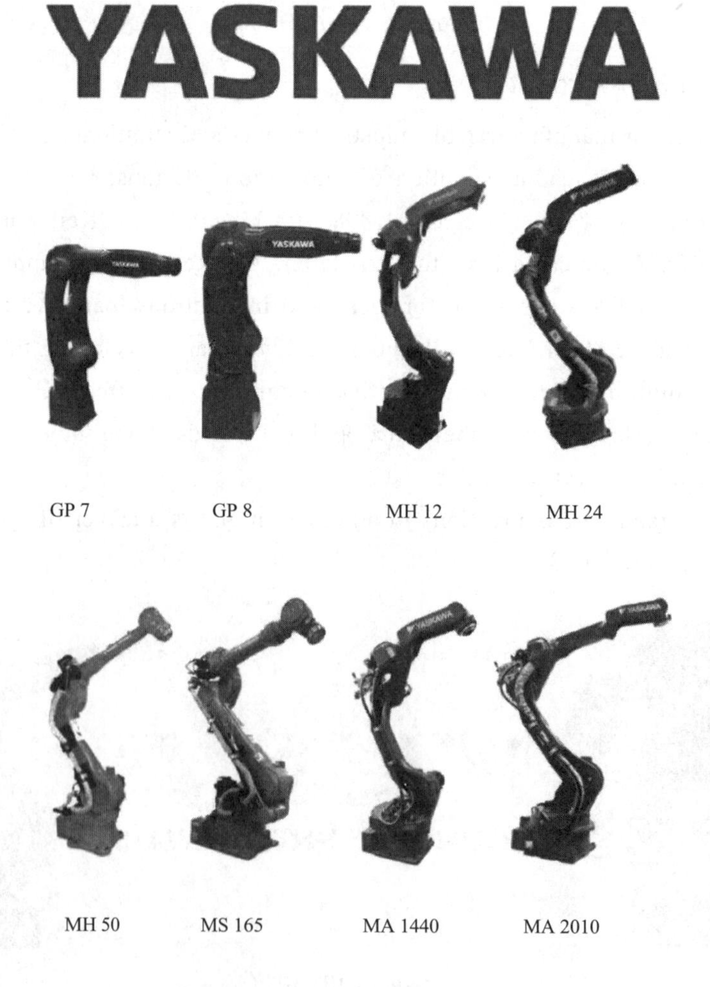

Figure 14.19 YASKAWA robots

4) Others (see Figure 14.20 and Figure 14.21)

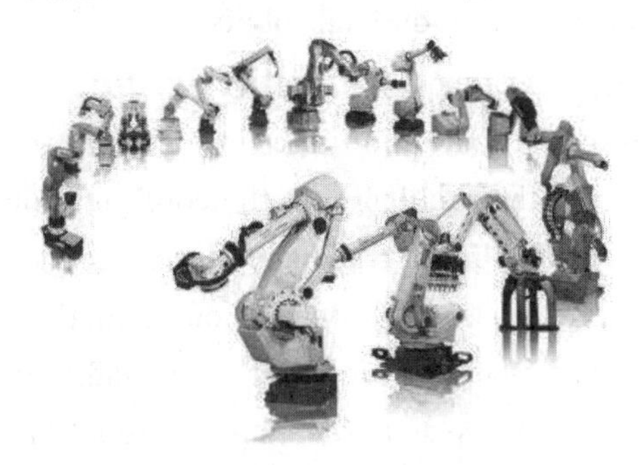

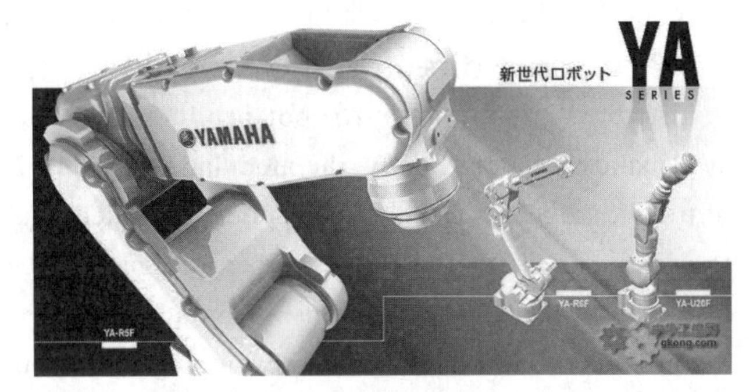

Figure 14.20 FANUC

Figure 14.21 YAMAHA

Extension Materials

Development Trend of Robot

1. Robot and information technology in-depth integration

Big data and cloud storage technology make the robot gradually become the terminal and node of the Internet of things. Firstly, the rapid development of information technology will be the integration of industrial robots and networks to form a complex production system. A variety of algorithms such as ant colony algorithm, immune algorithm can be gradually applied to the application of robots, so that robots have people's ability to learn. Multiple robots collaboration technology makes a set of production solutions possible. Secondly, the service robot is generally able to achieve remote monitoring through network. Multiple robots can provide more processes, more complex operations of the service; the new mode of operation which human awareness can control robot is also being developed, that is, the use of "thinking" and "willpower" controls the behavior of the robot.

2. Promote ease of use and stability of robotic products

With the technologies of standardization of robot structure, integration of joints, self-assembly and self-repair improved, the robot's ease of use and stability has been improved. Firstly, the application of robot has been extended to food, medical, chemical and other broader manufacturing areas, from the more mature cars, electronics industry; service areas and service objects of robots continue to increase; the body of robot has developed to small size, wide application. Secondly, the cost of the robot has developed. Robot technology and process are becoming more mature. Compared with traditional special equipment, the initial price gap of robots has been narrow, and in a high degree of personalization, cumbersome process, the robots have higher economic efficiency. Thirdly, man-machine relationship has undergone profound changes. For example, when workers and robots work together, robot can understand human language, graphics, and physical commands through a simple way, eliminate complex worker operations using modular plugs and production components. There is a big safety problem in man-machine collaboration at the existing stage, although light industrial robots with visual and advanced sensors have been developed, there is still a lack of reliable technical specifications for industrial robots.

3. Robots direct to the modular, intelligent and systematic way

At present, the global robot products have developed to the modular, intelligent and systematic direction. Firstly, the modular design has improved the problem that the configuration of traditional robots can only be applied to limited range, industrial robot research has more tended to use reconstruction, modular product design ideas, which help users to solve the contradictions of product variety, specifications, manufacturing cycle with the cost of production. Secondly, in the process of the robot product developing intelligent, the control system of industrial robot has developed to the direction of the open control system integration, the servo drive technology has changed to the direction of the unstructured, multi-mobile robot system, robot collaboration has been not only the coordination of control, but been the coordination of the organization and control of the robot system. Thirdly, the industrial robot technology continues to extend, the current robot products are being embedded to construction machinery, food machinery, laboratory equipment, medical equipment and other traditional equipment.

4. The market demand for new intelligent robots increases

The demand for new intelligent robots, especially those with intelligence, flexibility, cooperation and adaptability, continues to grow. Firstly, the fine work ability of the next generation of intelligent robots is further enhanced; the ability to get adapted to the outside world has been increasing. In terms of the fine operating capability of the robot, the Boston Consulting Group survey showed that robots entering factories and laboratories recently had distinct characteristics that could accomplish fine work such as assembling tiny parts; pre-programmed robots don't need expert monitoring. Secondly, the market demand for robot flexibility continues to increase. Recently, Renault used a batch of screwing robot of which the load is 29 kg, the only 1.3 m long arm in which there are six rotary joints embedded can be flexible operation. Thirdly, the demand of interpersonal skills continues to grow. In future, robots are able to perform tasks near workers. A new generation of intelligent robots use sonar, cameras, or other technologies to sense if there are workers in the work environment, and if there is a collision possible, they may slow down or stop functioning.

机器人的发展趋势

1. 机器人与信息技术深入融合

大数据和云存储技术使得机器人逐步成为物联网的终端和节点。一是信息技术的快速发展将工业机器人与网络融合，组成复杂性强的生产系统。各种算法，如蚁群算法、免疫算法等可以逐步应用于机器人中，使其具有类人的学习能力。多台机器人协同技术使一套生产解决方案成为可能。二是服务机器人普遍能够通过网络实现远程监控，多台机器人能提供流程更多、操作更复杂的服务；人类意识控制机器人这一新操作模式也正在研发中，即利用“思维力”和“意志力”控制机器人的行为。

2. 机器人产品易用性与稳定性提升

随着机器人标准化结构、集成一体化关节、自组装与自修复等技术的改善，机器人的易用性与稳定性不断提高。一是机器人的应用领域已经从较为成熟的汽车、电子产业延伸至食品、医疗、化工等更广泛的制造领域；服务领域和服务对象不断增加，机器人本体小、应用广。二是机器人成本快速下降。机器人技术和工艺日趋成熟。机器人初期投资相较于传统专用设备的价格差距缩小，在个性化程度高、工艺和流程烦琐的产品制造中具有更高的经济效益。三是人机关系发生了深刻改变。例如，工人和机器人共同工作时，机器人能够通过简易的感应方式理解人类语言、图形、身体指令，利用其模块化的插头和生产组件，免除工人复杂的操作。现有阶段的人机协作存在较大的安全问题，尽管具有视觉和先进传感器的轻型工业机器人已经被开发出来，但是目前仍然缺乏安全可靠的工业机器人协作的技术规范。

3. 机器人向模块化、智能化和系统化方向发展

目前全球推出的机器人产品向模块化、智能化和系统化方向发展。第一，模块化改变了传统机器人结构仅能适应有限范围的问题，工业机器人的研发更趋向采用组合式、模块化的产品设计思路，重构模块化帮助用户解决产品品种、规格与设计制造周期和生产成本之间的矛盾。第二，机器人产品向智能化发展的过程中，工业机器人控制系统向开放性控制系统集成方向发展，伺服驱动技术向非结构化、多移动机器人系统改变，机器人协作已经不仅是控制的协调，而且是机器人系统的组织与控制方式的协调。第三，工业机器人技术不断延伸，目前的机器人产品正嵌入工程机械、实验设备、医疗器械等传统装备之中。

4. 新型智能机器人市场需求增加

新型智能机器人，尤其是具有智能性、灵活性、合作性和适应性的机器人的需求持续增长。第一，下一代智能机器人的精细作业能力进一步提升对外界的感知能力不断增强。在机器人精细作业能力方面，波士顿咨询集团调查显示，最近进入工厂和实验室的机器人具有明显不同的特质，它们能够完成精细化的工作内容，如组装微小的零部件，预先设定程序的机器人不再需要专家的监控。第二，市场对机器人灵活性方面的需求不断提高。雷诺目前使用了一批 29 kg 的拧螺钉机器人，它们在仅有的 1.3 m 长的机械臂中嵌入六个旋转接头，机械臂能灵活操作。第三，机器人与人协作能力的要求不断增强。未来机器人能够靠近工人执行任务，新一代智能机器人采用声呐、摄像头或者其他技术感知工作环境是否有人，如有碰撞可能，它们会减慢速度或者停止运行。

Self-Test

Complete the following topics according to what you have learned.

(1) “Painting robot” is an industry term for a robot that has two major differences from all

other standard industrial robots. ( ) ( ).

(2) Robots can be made from a variety of materials including metals and plastics. The industrial robot is composed of d parts: ( ), ( ), ( ) and ( ).

New Words and Phrase

robot *n.* 机器人
machine *n.* 机器
v. 用机器制造
labor *n.* 劳动，工作，努力
factory *n.* 工厂，制造厂
describe *v.* 描述，描写
efficient *adj.* 有效率的，有能力的
emotionless *adj.* 没有情感的，冷漠的
incapable *adj.* 不能的，不能胜任的
engineering *n.* 工程，工程学
mechanics *n.* 力学，结构，机械学
behavior *n.* 行为，举止，态度，反应
cognition *n.* 认识，感知
autonomy *n.* 自治，自治权
injure *v.* 伤害，损害
industrial robot 工业机器人
teaching box 示教盒
robotic hand 机器手
rigid body 刚体
external force 外力
rotary joint 旋转接头，旋转关节
pin joint 针接头
hinge joint 链接头
kinematic pair 运动副
relative movement 相对运动
articulated robot 较接式机器人
continuous path 连续路径
motion control 运动控制
direct drive 直接驱动
given point 给定点
folding mechanism 折叠机构
uni-axial rotation 单轴旋转
control scheme 控制方案

CHAPTER 6

Expansion

The points and criteria of translation about professional English:

(1)More long and complicated sentences: Technology articles require accurate description, reasoning, and complicated sentence. The translation should be appropriate to use clause to describe. So the articles can adjust clearly and they are easy to be understand.

(2)More passive voice: Chinese active sentences and passive sentences can be used.

(3)More non-finite verb: Each simple sentence in English can only use one verb. If there are more verbs, you must select the main verb. Other look as a non-finite verb from that can conform the requirement of the grammar.

(4)More word transform: You should be flexible while translating. You must know the word clearly to have a correct understanding of the full sentences.

The translation standards of professional English are accurate, clear and easy to understand.

Unit 15

PID Controllers

In Figure 15.1, it is shown the application of water level PID control system. In Figure 15.2, it is shown the PID parameter setting interface of control system . From the following passage you will learn more about PID control system.

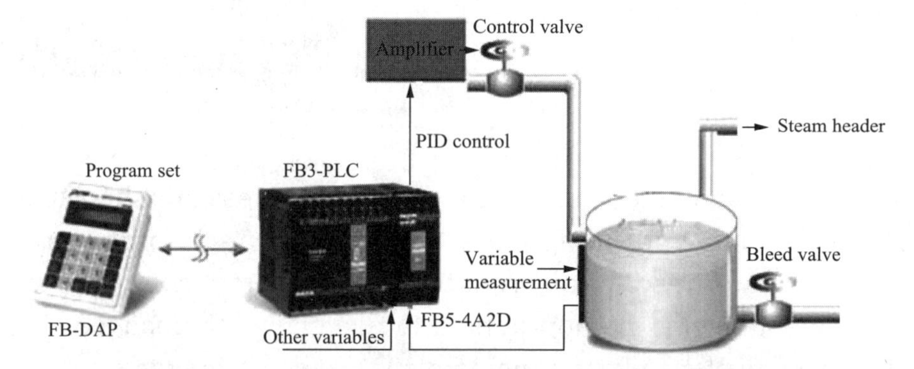

Figure 15.1 Application of water level PID control system

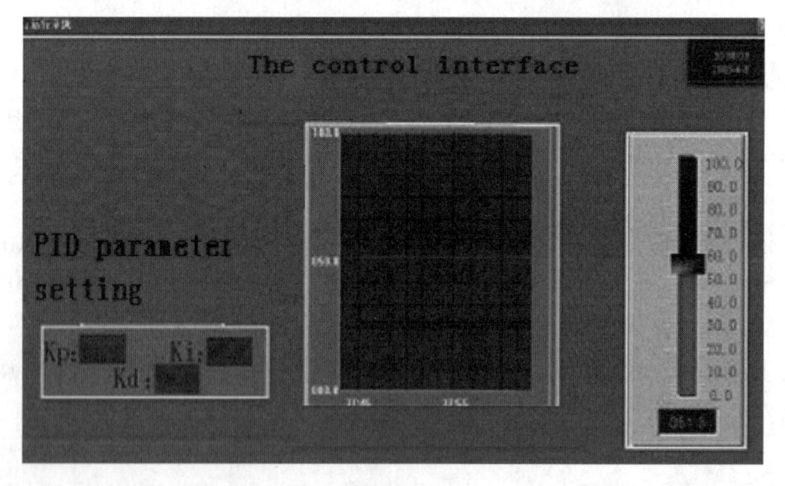

Figure 15.2 The PID parameter setting interface of control system

PID controllers can be stand-alone controllers (also called single loop controllers), controllers in PLCs, embedded controllers, or software in Visual Basic or C# computer programs.

PID controllers are process controllers with the following characteristics:

(1) Continuous process control.

(2) Analog input (also known as "measurement" or "process variable" or "PV").

(3) Analog output (referred to simply as "output").

(4) Set point (SP).

(5) Proportional (P) , integral (I) , and/or derivative (D) constants.

Examples of "continuous process control" are temperature, pressure, flow, and level control. For example, control the heating of a tank. For simple control, you have two temperature limit sensors (one low and one high) and then switch the heater on when the low temperature limit sensor turns on and then turn the heater off when the temperature rises to the high temperature limit sensor. This is similar to most home air conditioning and heating thermostats.

In contrast, the PID controller would receive input as the actual temperature and control a valve that regulates the flow of gas to the heater. The PID controller automatically finds the correct (constant) flow of gas to the heater that keeps the temperature steady at the set point. Instead of the temperature bouncing back and forth between two points, the temperature is held steady. If the set point is lowered, then the PID controller automatically reduces the amount of gas flowing to the heater. If the set point is raised, then the PID controller automatically increases the amount of gas flowing to the heater. Likewise the PID controller would automatically for hot, sunny days (when it is hotter outside the heater) and for cold, cloudy days.

The analog input (measurement) is called the "process variable" or "PV". You want the PV to be a highly accurate indication of the process parameter you are trying to control. For example, if you want to maintain a temperature of + or − one degree Celsius then we typically strive for at least ten times that or one-tenth of a Celsius. If the analog input is a 12-bit analog input and the temperature range for the sensor is 0 to 400 Celsius then our "theoretical" accuracy is calculated to be 400 Celsius divided by 4,096 (12 bits) = 0.097,656,25 Celsius [1]. We say "theoretical" because it would assume there was no noise and error in our temperature sensor, wiring, and analog converter. There are other assumptions such as linearity, etc. The point being—with 1/10 of a degree Celsius "theoretical" accuracy—even with the usual amount of noise and other problems one degree of accuracy should easily be attainable.

The analog output is often simply referred to as "output". Often this is given as 0%~100%. In this heating example, it would mean the valve is totally closed (0%) or totally open (100%) .

The set point (SP) is simply—what process value do you want. In this example—what temperature do you want the process at?

The PID controller's job is to maintain the output at a level so that there is no difference (error) between the process variable (PV) and the set point (SP) .

In Figure 15.3, the valve could be controlling the gas going to a heater, the chilling of a cooler, the pressure in a pipe, the flow through a pipe, the level in a tank, or any other process control system.

What the PID controller is looking at is the difference (or "error") between the PV and the SP. It looks at the absolute error and the rate of change of error. Absolute error means—is there a big difference in the PV and SP or a little difference? Rate of change of error means—is the difference between the PV or SP getting smaller or larger as time goes on.

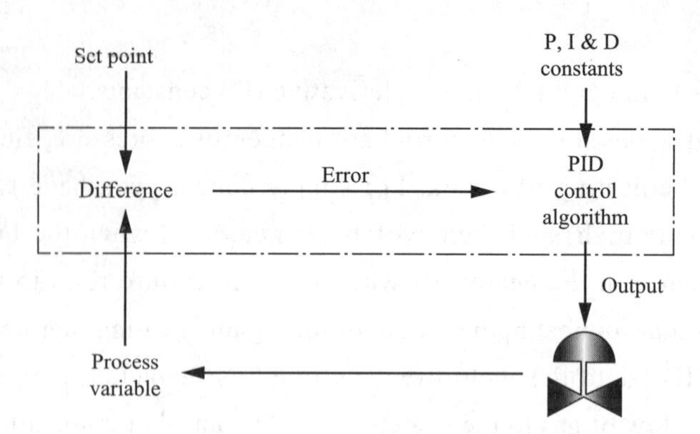

Figure 15.3 PID controller

When there is a "process upset", meaning, when the process variable or the set point quickly changes—the PID controller has to quickly change the output to get the process variable back equal to the set point[2]. If you have a walk-in cooler with a PID controller and someone opens the door and walks in, the temperature (process variable) could rise very quickly. Therefore the PID controller has to increase the cooling (output) to compensate for this rise in temperature.

Once the PID controller has the process variable equal to the set point, a good PID controller will not vary the output. You want the output to be very steady (not changing) . If the valve (motor, or other control element) is constantly changing, instead of maintaining a constant value, this could cause more wear on the control element.

So there are these two contradictory goals. Fast response (fast change in output) when there is a "process upset", but slow response (steady output) when the PV is close to the set point.

Note that the output often goes past (over shoots) the steady-state output to get the process back to the set point. For example, a cooler may normally have its cooling valve open 34% to maintain zero Celsius (after the cooler has been closed up and the temperature settled down). If someone opens the cooler, walks in, walks around to find something, then walks back out, and then closes the cooler door—the PID controller is breaking out because the temperature may have raised 20 Celsius! So it may crank the cooling valve open to 50%, 75%, or even 100%—to hurry up and cool the cooler back down—before slowly closing the cooling valve back down to 34%[3]. Let's think about how to design a PID controller.

We focus on the difference (error) between the process variable (PV) and the set point (SP). There are three ways we can view the error.

1. The absolute error

This means how big is the difference between the PV and SP. If there is a small difference between the PV and the SP—then let's make a small change in the output. If there is a large difference in the PV and SP—then let's make a large change in the output. Absolute error is the "proportional" (P) component of the PID controller.

2. The sum of errors over time

Give us a minute and we will show why simply looking at the absolute error (proportional) only is a problem. The sum of errors over time is important and is called the "integral" (I) component of the PID controller. Every time we run the PID algorithm we add the latest error to the sum of errors. In other words Sum of Errors = Error1 + Error2 + Error3 + Error4 +⋯.

3. The dead time

Dead time refers to the delay between making a change in the output and seeing the change reflected in the PV. The classical example is getting your oven at the right temperature. When you first turn on the heat, it takes a while for the oven to "heat up". This is the dead time. If you set an initial temperature, wait for the oven to reach the initial temperature, and then you determine that you set the wrong temperature – then it will take a while for the oven to reach the new temperature set point. This is also referred to as the "derivative" (D) component of the PID controller. This holds some future changes back because the changes in the output have been made but are not reflected in the process variable yet.

Absolute error/proportional. One of the first ideas people usually have about designing an automatic process controller is what we call "proportional". Meaning, if the difference between the PV and SP is small – then let's make a small correction to the output. If the difference between the PV and SP is large—then let's make a larger correction to the output. This idea certainly makes sense.

We simulated a proportional only controller in Microsoft Excel. Figure 15.4 is the chart showing the results of the first simulation (DEADTIME=0, proportional only).

Proportional and integral controllers. The integral portion of the PID controller accounts for the offset problem in a proportional only controller. We have another Excel spreadsheet that simulates a PID controller with proportional and integral control. Here (see Figure 15.5) is a chart of the first simulation with proportional and integral (DEADTIME = 0, proportional = 0.4).

As you can tell, the PI controller is much better than just the P controller. However, dead time of zero is not common.

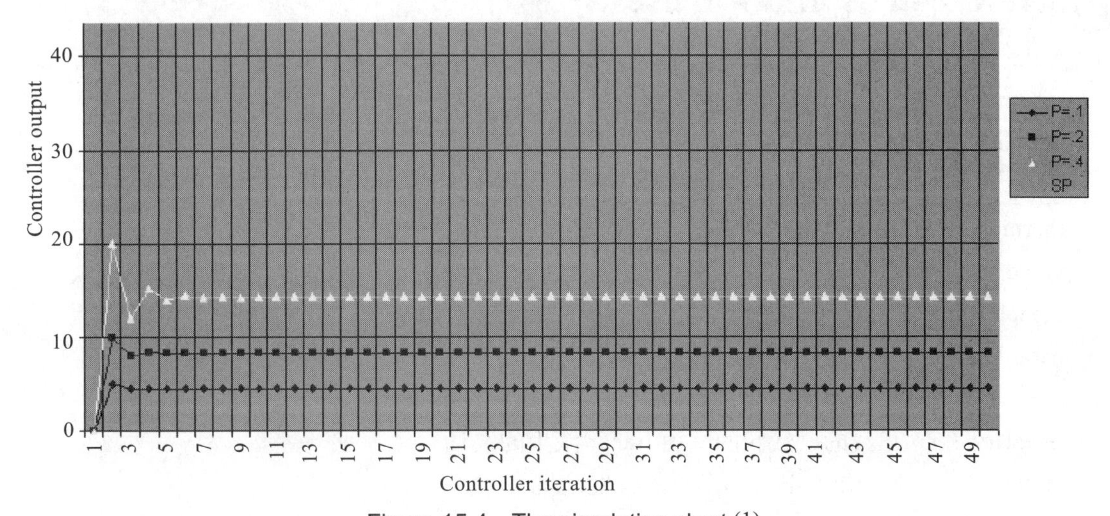

Figure 15.4 The simulation chart (1)

Derivative control. Derivative control takes into consideration that if you change the output, then it takes time for that change to be reflected in the input (PV) [4]. For example, let's take heating of the oven.

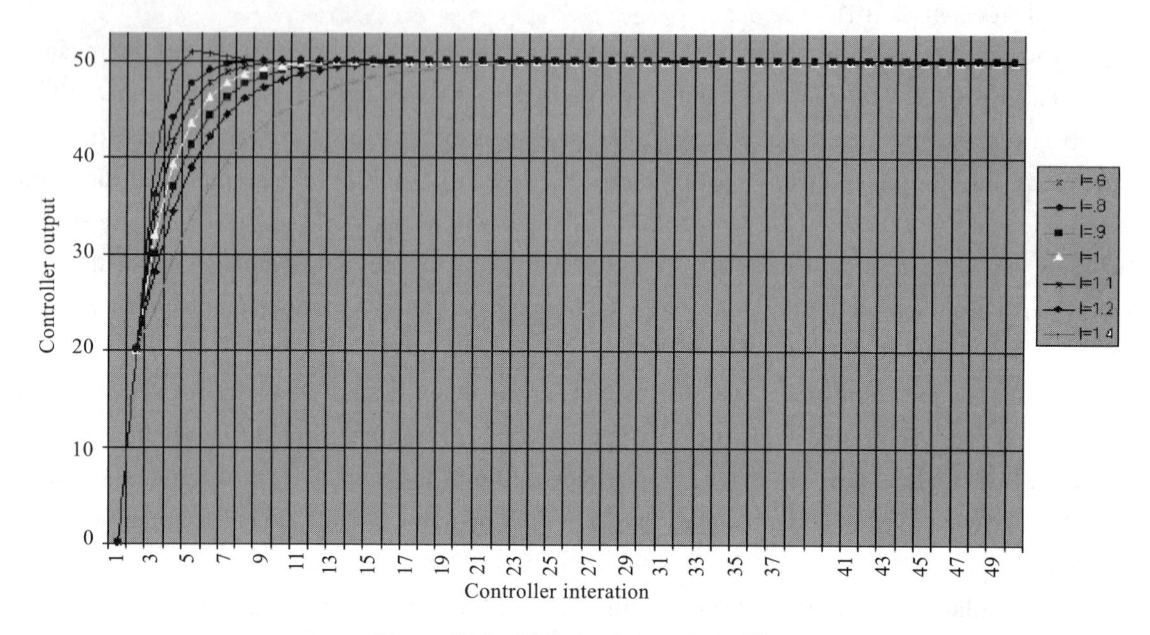

Figure 15.5 The simulation chart (2)

If we start turning up the gas flow, it will take time for the heat to be produced, the heat to flow around the oven, and for the temperature sensor to detect the increased heat. Derivative control sort of "holds back" the PID controller because some increase in temperature will occur without needing to increase the output further. Setting the derivative constant correctly allows you to become more aggressive with the P and I constants.

New Words and Phrases

make sense 有意义，有道理；可实行的

integral *adj.* 完整的，整数的，积分的

derivative *adj.* 衍生的

n. 派生词，衍生物

thermostat *n.* 温度自动调节器，恒温器

valve *n.* 阀

compensate *v.* 补偿，赔偿；抵消，弥补

parameter *n.* 参数（量）

strive *v.* 努力，力求

linearity *adj.* 线的，线状的（由 linear 派生的名词）

motor *n.* （尤指使用电或内燃机的）发动机

adj. 起动的，发动的；用发动机的

element *n.* 成分，因素，元素；电阻丝

contradictory *adj.* 相互矛盾的，相互抵触的

freak *n.* 反常的事，不正常的人

crank *n.* （可将往复运动与循环运动互为转换的）曲柄，曲轴

v. 用曲柄启动（转动）

algorithm *n.* 算法；计算程序

oven *n.* 烤箱，烤炉

initial *adj.* 开始的，最初的

component *n.* 组成部分，零件

adj. 组成的，构成的

correction *n.* 纠正，校正

simulate *v.* 模仿，模拟

offset *n./v.* 抵消，补偿

account for 说明，解释，构成

Analyze the Following Sentences

[1] If the analog input is a 12-bit analog input and the temperature range for the sensor is 0 to 400 Celsius then our "theoretical" accuracy is calculated to be 400 Celsius divided by 4,096 (12 bits) = 0.097,656,25 Celsius.

如果是一个 12 位的模拟输入，传感器的温度范围是 0~400℃，我们计算的理论精确度就是 4 096（12 位）÷400 ℃ = 0.097 656 ℃。（注：这是一个条件从句，"for the sensor"是状语）

[2] When there is a "process upset", meaning, when the process variable or the set point quickly changes − the PID controller has to quickly change the output to get the process variable back equal to the set point.

如果存在过程扰动，即过程变量或基准点变化时——PID 控制器就要迅速改变输出，这样过程变量就返回到基准点。（注：这是一个由 when 引导的时间状语从句，"meaning"引导的是并列关系的从句，"to get the process variable back equal to the set point"是目的状语从句。）

[3] So it may crank the cooling valve open to 50%, 75%, or even 100%—to hurry up and cool the cooler back down—before slowly closing the cooling valve back down to 34%.

这样制冷阀门就可能打开 50%、75% 甚至 100%，目的是在慢慢关闭制冷阀门到它的 34% 之前，赶快降低制冷器的温度。（注：句中"to hurry up and cool the cooler back down"是解释说明部分。）

[4] Derivative control takes into consideration that if you change the output, then it takes time for that change to be reflected in the input (PV) .

微分控制考虑的是：如果你改变输出，那么要在输入（PV）处反映这个改变就需要些时间。（注：句中"takes into consideration"是"考虑到，斟酌"的意思，"that if you change"引导的是宾语从句，"to be reflected in the input"是省略引导词的定语从句，修饰"change"。）

Translation

图 15.1 所示的是 PID 在水位控制系统中的应用；图 15.2 所示的是控制系统的 PID 参数设定界面。下面的内容将会使你了解更多关于 PID 控制系统的知识。

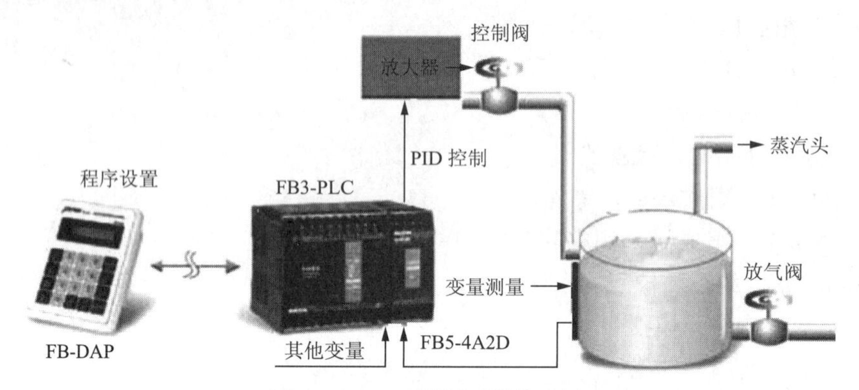

图 15.1 PID 控制系统的应用

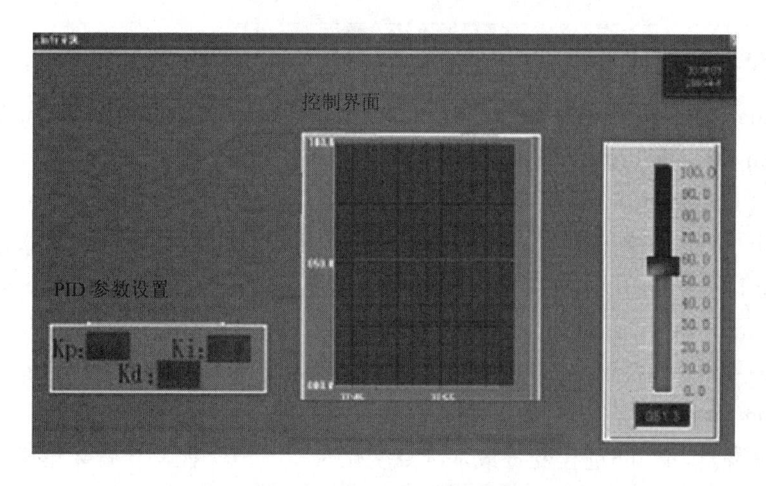

图 15.2 PID 控制系统的控制界面

PID 控制器可以是独立控制器（又称单回路控制器），可编程控制器（PLC）中的控制器，嵌入式控制器或者是用 VB 或 C# 编写的计算机程序软件。

PID 控制器是过程控制器，它具有如下特征：

（1）连续过程控制。

（2）模拟输入（又称“测量量”或“过程变量”或“PV”）。

（3）模拟输出（简称“输出”）。

（4）基准点（SP）。

（5）比例、积分以及 / 或者微分常数。

“连续过程控制”的例子有温度、压力、流量及水位控制。例如。控制一个容器的热量。对于简单的控制，可使用两个具有温度限定功能的传感器（一个限定低温，一个限定高温）。当低温限定传感器接通时就会打开加热器；当温度升高到高温限定传感器时就会关闭加热器。这类似于大多数家庭使用的空调及供暖系统的温度自动调节器。

反过来，PID 控制器能够接受像实际温度这样的输入，控制阀门，这个阀门能够控制进入加热器的气体流量。PID 控制器自动地找到加热器中气体的合适流量，这样就保持了温度在基准点稳定。温度稳定了，就不会在高低两点间上下跳动了。如果基准点降低，PID 控制器就会自动降低加热器中气体的流量；如果基准点升高，PID 控制器就会自动增加加热器中气体的流量。同样地，对于高温，晴朗的天气（当外界温度高于加热器温度时）及阴冷，多云的天气，PID 控制器都会自动调节。

模拟输入（测量量）又称“过程变量”或“PV”。一般希望 PV 能够达到所控制过程参数的高精确度。例如，如果想要保持温度为 +1℃或 –1℃，我们至少要为此努力，使其精度保持在 0.1 度。如果是一个 12 位的模拟输入，传感器的温度范围是 0℃到 400℃，我们计算的理论精确度就是 4 096 ÷400℃ =0.097 656 25℃。之所以说这是理论，是因为假定温度传感器、导线及模拟转换器上没有噪声和误差，还有其他的假定。例如，线性等。即使是有大量的噪声和其他问题，按理论精确度的 1/10 计算，1℃精确度的数值应该是很容易得到的。

模拟输出经常简称“输出”。数值经常为 0%~100%。在这个例子中，阀门完全关闭为 0%，完全打开为 100%。

基准点(SP)很简单,即想要什么样的过程量。在这个例子中,即想要过程处于怎样的温度。PID 控制器的任务是维持输出在一个程度上，这样在过程变量（PV）和基准点（SP）上就没有偏差（误差）。

在图 15.3 中，阀门用来控制进入加热器的气体，冷却器的制冷，水管的压力，水管的流量，容器的水位或其他的过程控制。

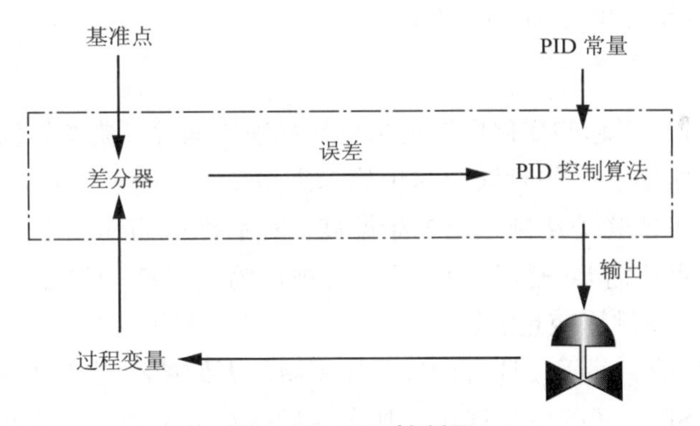

图 15.3　PID 控制器

PID 控制器所观察的是 PV 和 SP 之间的偏差。它观察绝对偏差和偏差变换率。绝对偏差就是 PV 和 SP 之间偏差大还是小。偏差变换率就是 PV 和 SP 之间的偏差随着时间的变化是越来越小还是越来越大。

如果存在过程扰动，即过程变量或基准点变化时，PID 控制器就要迅速改变输出，这样过程变量就返回到基准点。如果你有一个 PID 控制的可进入的冷冻装置，某个人打开门进入，温度（过程变量）将会迅速升高。因此，PID 控制器不得不提高冷度（输出）来补偿这个温度的升高。

一旦过程变量等同于基准点，一个好的 PID 控制器就不会改变输出。你所要的输出就会稳定（不改变）。如果阀门（发动机或其他控制元件）不断改变，而不是维持恒量，这将造成控制元件更多的磨损。

这样就有了两个矛盾的目标。当有“过程扰动”时能够快速反应（快速改变输出）。当 PV 接近基准点时就缓慢反应（平稳输出）。

我们注意到，输出量经常超过稳定状态，输出使过程变量回到基准点。比如，一个制冷器通常打开它的制冷阀门的 34%，就可以维持在 0 ℃（在制冷器关闭和温度降低后）。如果有人打开冷却器，到处走走，找点东西，然后再走回去，再关上制冷器的门，PID 控制器会非常活跃，因为温度可能将上升 20 ℃。这样制冷阀门就可能打开 50%、75% 甚至 100%，目的是在慢慢关闭制冷阀门到它的 34% 之前，赶快降低制冷器的温度。

让我们思考一下如何设计一个 PID 控制器。

我们主要集中在过程变量（PV）和基准点（SP）之间的偏差（误差）上。有三种定义误差的方式。

1. 绝对偏差

这说明的是 PV 和 SP 之间的偏差有多大。如果 PV 和 SP 之间偏差小，那我们就在输出时做一个小的改变。如果 PV 和 SP 之间偏差大。那我们就在输出时做一个大的改变。绝对偏差就是 PID 控制器的比例环节。

2. 累积误差

请给我们几分钟，然后我们将说明为什么不能只查看绝对误差（比例环节）。累积误差是很重要的，我们把它称为是 PID 控制器的积分环节。每次我们运行 PID 算法时，我们总会把最近的误差添加到误差总和中。换句话说，累积误差 = 误差 1 + 误差 2 + 误差 3 + 误差 4 + …。

3. 死区时间

死区时间指的是 PV 引起的变化由发现到改变之间的延时。典型的例子就是调整你的烤炉在合适的温度。当你刚刚加热的时候，烤炉热起来需要一定时间。这就是滞后时间。如果你设置一个初始温度，等待烤炉达到这个初始温度，然后你认为你设定了错误的温度，烤炉达到这个新的温度基准点还需要一段时间。这也就被认为是 PID 控制器的微分环节，可以抑制了未来的改变，但并不受过程变量的影响。

绝对偏差 / 比例环节。关于设计自动过程控制器，人们最初想法之一是设计比例环节。意思就是，如果 PV 和 SP 之间的偏差很小，那么我们就在输出处做一个小的修改；如果 PV 和 SP 之间的偏差很大，那么我们就在输出处做一个大的修改。当然这个想法是有意义的。

我们在 Microsoft Excel 仅对比例控制器进行仿真。图 15.4 是显示首次仿真结果的表格（滞后时间 = 0，只含比例环节）。

比例、积分控制器。PID 控制器中的积分环节是用来负责纯比例控制器中的补偿问题的。我们有另外一个 Excel 的扩展表格，表格上仿真的是一个具有比例积分功能的 PID 控制器。这里（见图 15.5）是比例积分控制器最初的仿真表格（滞后时间 = 0，比例常数 = 0.4）。

众所周知，比例积分控制器要比仅有比例功能的比例控制器好得多，但是等于 0 的滞后时间并不常见。

微分控制。微分控制考虑的是：如果你改变输出，那么要在输入（PV）处反映这个改变就需要些时间。比如，以烤炉的加热为例。

如果我们增大气体的流量，那么从产生热量，热量分布烤炉的四周，到温度传感器检测升高的温度都将需要时间。PID 控制器中微分环节具有抑制功能，因为有些温度增量会在以后不需要的情况下产生。正确地设置微分常数有利于对比例常数和积分常数的确定。

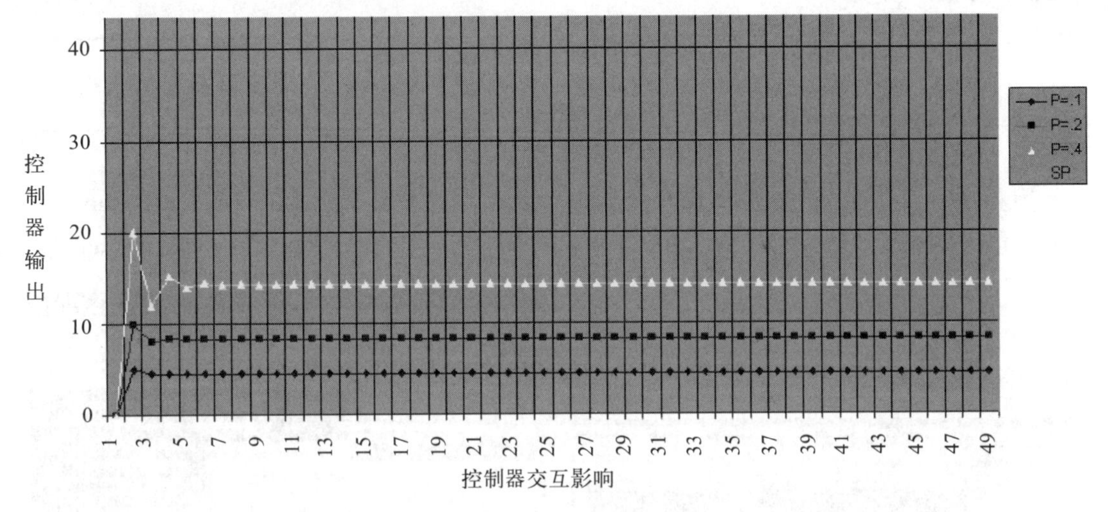

图 15.4 仿真曲线（1）

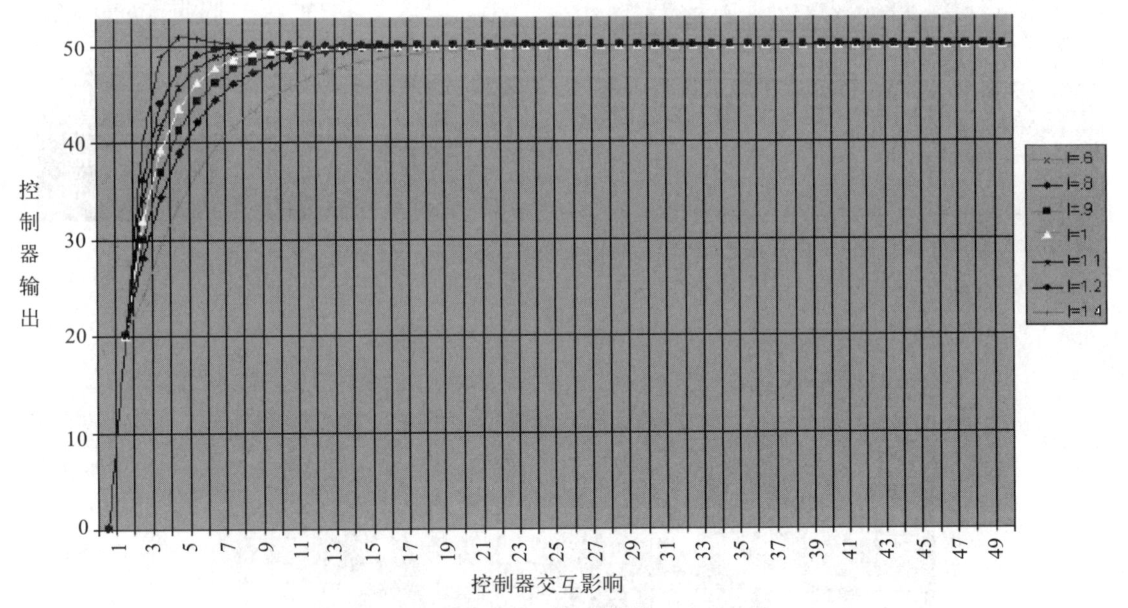

图 15.5 仿真曲线（2）

Unit 16

CAD and CAM

In Figure 16.1, it is shown the operation interface of CAM. In Figure 16.2, it is shown the operation interface of CAD. From the following passage you will learn more about CAM and CAD.

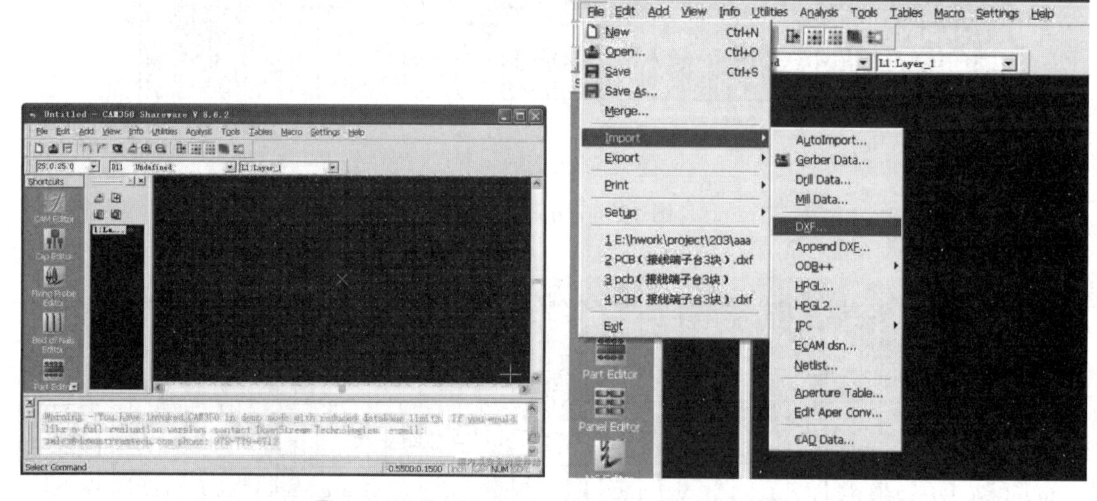

Figure 16.1 The operation interface of CAM

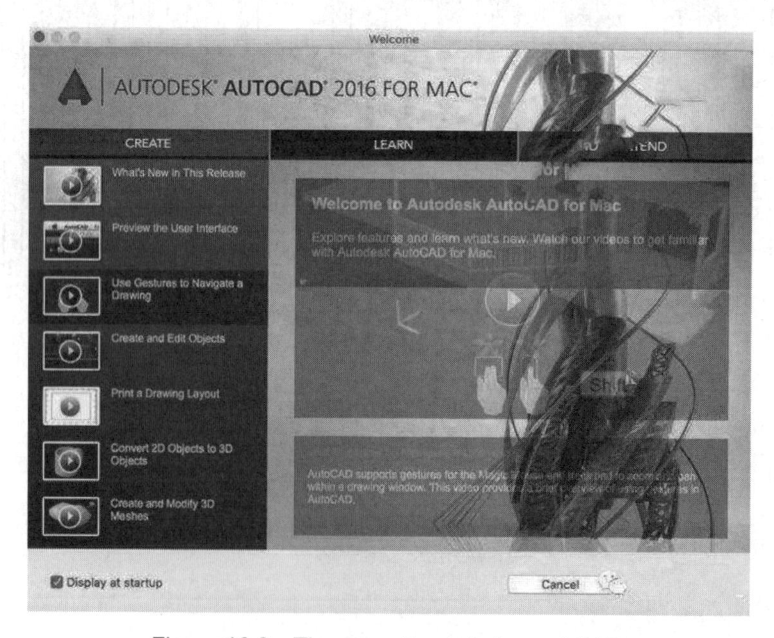

Figure 16.2 The operation interface of CAD

The term CAD/CAM is a shortening of computer-aided design (CAD) and computer-aided manufacturing (CAM). Well then, what is a general CAD system?

The general CAD system was developed by considering a wide range of possible uses of such a system. The following were considered in detail:

(1) Mechanical engineering design;

(2) Building design;

(3) Structural engineering design;

(4) Electronic circuit design;

(5) Animation and graphic design.

It was postulated that four basic processes involving graphics occurred, to various degrees, in each field, namely:

(1) Pure analysis—standard design and analysis processes.

(2) Pure draughting—production of a drawing or picture by the manual creation and manipulation of lines, arcs, etc.

(3) Drawing by analysis—the production of a picture or part of a picture directly from analysis: for example, production of cam profiles.

(4) Analysis of drawing—evaluation of the properties of an item described graphically, for example the production of a quantity list by analysis of a builder's plan drawing.

For the system to be able to support pure analysis it must contain facilities for the running of analysis programs of unlimited length and for the storage and rapid retrieval of large amounts of data.

It was considered important that the user should be able to communicate directly and graphically with analysis programs. Graphics facilities were provided which were considered to be sufficient for a general design draughting system. However, the range of graphical construction techniques is so large in practice that the system contained only as many facilities as could practically be incorporated in the draughting system, leaving other more specialized techniques to be developed by the applications programmer[1].

For both the production of drawing items by analysis and the analysis of drawings, it is essential that there is a simple efficient link between data produced by the draughting system and analysis programs. It is also essential that graphic data can be annotated in a way which is recognized by analysis programs but which does not affect the draughting system[2].

It was thought that for most practical application the general draughting system would be incorporated in a much larger specific applications system. For this reason the draughting system was as simple as possible consistent with reasonable running efficiency, so that it could be incorporated into an applications system with the minimum of effort.

The facilities embodied in the general CAD system are now described. These facilities are aimed at allowing a user to input graphical information into the computer and file it. Initial data entry is made by digitizing rough sketches. The system also permits the user to access the data, manipulate it, process it, output it in hard-copy form, or re-file it for permanent storage.

There are many reasons for using CAD; the most potent driving force is competition. In order to win business, companies used CAD to produce better designs more quickly and more cheaply than their competitors. Productivity is much improved by a CAD program enabling you to easily draw polygons, ellipses, multiple parallel lines and multiple parallel curves. Copy, rotate and mirror facilities are also very handy when drawing symmetrical parts. Many hatch patterns are supplied with CAD programs. Filling areas in various colors is a requirement in artwork and presentations. Different style fonts for text are always supplied with any CAD programs. The possibility of importing different graphic file formats and scanning of material (photographs) into a CAD program is also an asset especially as the image can be manipulated, retouched and animated.

Another advantage of a CAD system is its ability to store entities, which are frequently used on drawings. Libraries of regularly used parts can be purchased separately or can be created by the draughtsman. For repetitive use on a drawing, a typical item may be retrieved and positioned in seconds, also oriented at any angle to suit particular circumstances.

Using CAD products, assembly drawings can be constructed by inserting existing component drawings into the assembly drawing and positioning them as required.

Clearance between different components can be measured directly from the drawing, and if required, additional components designed using the assembly as reference.

CAD is very suitable for fast documentation. Previously, engineers and drafters wasted almost 30% of their time looking for drawings and other documents. Editing drawings to effect revisions and produce updated parts lists is quick and easy using a CAD product.

When you're working on paper and a customer wants to change a drawing, you have to draw it all over again; In CAD, you make the change immediately and print out a new drawing in minutes, or you can transmit it via e-mail or Internet all over the world instantly. On paper creating complex geometry often involves a lot of measuring and location of reference points; In CAD it is a breeze and revisions are even simpler. Many CAD programs include a macro or an add-on programming language that allows customizing it.

Customizing your CAD programs to suit your specific needs and implementing your ideas can make your CAD system different from your rivals. CAD can enable companies to produce better designs that are almost impossible to produce manually and to eliminate dubious options during the conceptual design phase.

Many CAD systems permit the rapid generation of models of proposed designs as wireframes. The solid modeling created in CAD can be transferred to a finite element analysis (FEA) program, which will then verify whether the suggested design will be capable of supporting the expected loads[3].

CAD will be linked to CAM (computer aided manufacturing) whenever possible.

CAD/CAM systems could produce computerized instructions for computerized machine controllers: lathes, mills, machining centers, turret punches, welding equipment, automated assemblies, etc.

The CAM parts have evolved from the technology of numerical controlled (NC) machines. Early NC machines had their own on-board electronic control systems for their servo drives and motors, and where programmed by punched paper tape. In time, that becomes equivalent to a control stream of ASCII text data typed into a text editor.

Each machine maker developed their own control code scheme, usually a very cryptic set of letters for machine actions and numbers for the values of speed, depth, etc., and position coordinates.

NC machines include a computer with a screen and keyboard. These use a “conventional” control language. Modern CAD/CAM systems automatically generate tool paths from a 3D model, and can simulate the cutting action on-screen. The most CAD/CAM systems are modular, that means you can buy whichever modules do the option you want and they integrates into a unified system.

CIM (computer integrated manufacturing) means complete integration of all aspects of manufacturing utilizing computerized information.

CIM is the use of component data created by CAD in the CAM environment. In other words, the part geometry for manufacturing use in computerized form is used for NC programming. This stage of development may be termed small-scale integration.

The most highly developed form of CIM is the creation of a database containing all the information required for flexible manufacturing of components produced by the plant, in a form in which it can be retrieved and used by anyone who needs it. Flexible manufacturing means the ability to make any components in small numbers or well as large, quickly, at economical cost, thus reducing tool charges, work in process and costly inventory.

The main information flows involved in computer integrated manufacturing were clearly outlined by Helberg. CAD generates product model and product describing data that are transformed by CAPP (computer aided process planning) into routings and control programs for the CAM systems. The PPC (production planning and control) systems generate and manage all operational data that are used for controlling in the CAM area. CAQ (computer aided quality assurance) on a short-term basis corrects deviations in the manufacturing process and in the long run influences the development of products and methods with regard to quality assurance.

Helberg’s outline does not include further necessary or desirable informational connections between the systems, such as a connection of CAD/CAPP and PPC for an accompanying calculation during design and routing generation, or feedback from manufacturing to planning. Furthermore, at least in the case of single-parts manufacturing, processes like design and process planning can be regarded as elements of the lead time of an order and therefore can be planned and controlled by the PPC system in the same way as the actual manufacturing and assembly processed. [4] In that case a corresponding feedback becomes necessary.

System integration and rationalization is not simply a technological matter, as the CIM theorists suggest. To integrate disperse and incompatible systems we must change traditional procedure, not just throw in more money and equipment. Whenever we try to change procedures we find resistance. The larger the company and the more independent the network, the more difficult it is to turn policies and procedures around. Yet, as Figure 16.1 suggests, a condition for successful system integration is that it extends along functional and support lines, in the global sense of the distributed environment.

Figure 16.3 system integration should be accomplished along three different axes of reference: distributed environment (topology), functional support, software and hardware.

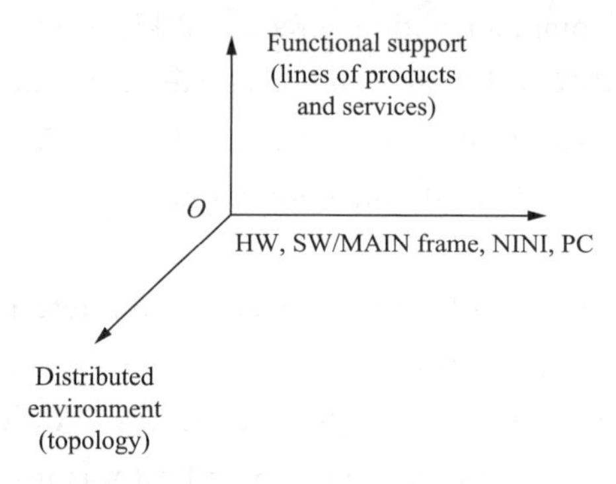

Figure 16.3 System integration

Because the tangible and intangible benefits of CIM are long term, the usual discounted cash-flow and return-on-investment methods cannot justify a CIM installation of a flexible manufacturing process frequently. Instead, strategic advantages and intangible benefits must be used to weigh the desirability of investment in CIM.

New Words and Phrases

postulate *vt.* 假定，假设

manipulation *n.* 操作，控制，处理，计算，运算

cam *n.* 凸轮

computer aided manufacturing 计算机辅助制造

computer-aided management 计算机辅助管理

computer-aided measurement 计算机辅助测量

retrieval *n.* 取回，索回；信息检索

annotate *vt. vi.* 注解，注释，评注

digitize *vt.* 将（资料）数字化

permanent *adj.* 永久性的，耐久的，固定不变的

potent *adj.* 有力的，有强效的，有势力的

polygon *n.* 多边形，多角形

ellipse *n.* 椭圆，椭圆形

symmetrical *adj.* 对称的，相称的
retouch *vt.* 修饰，润色，修改（绘图等）
revision *n.* 校订，订正，修改
breeze *n.* 微风，轻而易举的事
dubious *adj.* 可疑的，不确定的，含糊的，结果未定的，（工作等）无把握的
lathe *n.* 车床
vt. 用车床加工
mill *n.* 磨粉机，铣床，铣刀
punch *n.* 冲压机，冲床，打孔机
vt. 冲孔，打孔
weld *vt.* 焊接
n. 焊接，焊缝
servo *n.* 伺服，伺服系统
cryptic *adj.* 含义隐晦的
inventory *n.* 存货，库存量
disperse *vt.* 疏散，使散开，使分散
tangible *adj.* 切实的，可触知的，实在的，确实的
intangible *adj.* 难以明了的，无形的
CAPP (computer aided process planning) 计算机辅助工艺规划
PPC (production planning and control systems) 生产计划和控制系统
CAQ (computer aided quality assurance) 计算机辅助质量保证

Analyze the Following Sentences

[1] However, the range of graphical construction techniques is so large in practice that the system contained only as many facilities as could practically be incorporated in the draughting system, leaving other more specialized techniques to be developed by the applications programmer.

然而由于实际的制图技术范围太广，作图系统只能尽可能多地将实用工具包含进去，而将其他更专门的技术留给应用程序员去开发。（注：so large …that 后面是结果状语从句。）

[2] It is also essential that graphic data can be annotated in a way which is recognized by analysis programs but which does not affect the draughting system.

另外，图形数据可用能被分析程序识别但不对作图系统产生影响的方法进行注释。（注：定语从句中两个 which 的功能一样，都用于修饰 in a way。）

[3] The solid modeling created in CAD can be transferred to a finite element analysis (FEA) program, which will then verify whether the suggested design will be capable of supporting the expected loads.

在 CAD 里创建的实体造型可以输入到一个有限元分析（FEA）程序内，以检验设计方案能否承受预期的负荷。（注：定语从句中 which 修饰 FEA。）

[4] Furthermore, at least in the case of single-parts manufacturing, processes like design and process planning can be regarded as elements of the lead time of an order and therefore can be planned and controlled by the PPC system in the same way as the actual manufacturing and assembly processed. In that case a corresponding feedback becomes necessary.

此外，至少在加工单件零件的情况下，诸如设计和工艺规划这样的过程可以认为是订单交付周期的一部分。因此，它们可以像实际加工和装配过程一样由 PPC 系统规划和控制。在这种情况下，相应的反馈就变得非常必要。（注：the lead time of an order 可以翻译为"交付周期"）。

Translation

图 16.1 所示的是 CAM 的软件界面。图 16.2 所示的是 CAD 的软件界面。下面的内容将会使你了解更多关于 CAM 和 CAD 的知识。

CAD/CAM 是计算机辅助设计（CAD）和计算机辅助制造（CAM）的缩写。那么，什么是一个通用的 CAD 系统呢？

在开发通用 CAD 系统时要考虑该系统应具有尽可能广的应用范围，有以下几方面要详细考虑：

(1) 机械工程设计；

(2) 建筑设计；

(3) 结构工程设计；

(4) 电子电路设计；

(5) 动画和图形设计。

在各个领域，不论其应用程度如何，包括作图在内都应该具有四个基本过程，即：

(1) 纯分析——标准设计和分析过程。

(2) 纯绘图——用手工画线、圆弧等绘出的图或画。

(3) 分析作图——用直接分析的方法产生一幅图或图的一部分，例如作凸轮轮廓图。

(4) 绘图分析——对用图形方式描述的一个项目进行特性评估，如通过对建筑师规划图的分析，得出该项目的工程量表。

对于能支持纯分析的系统，必须提供运行无限长分析程序以及存储和快速检索大量数据的工具。

人们看重的一点是，用户能够通过直接和图形方式与分析程序通信。提供给通用绘图设计系统的作图工具应该充足。然而由于实际的制图技术范围太广，作图系统只能尽可能多地将实用工具包含进去，而将其他更专门的技术留给应用程序员去开发。

对于由分析产生的作图项目和绘图分析而言，由作图系统所产生的数据和分析程序之间必须有一个简单高效的联系。另外，图形数据可用能被分析程序识别但不对作图系统产生影响的方法进行注释。

对大多数实际应用来讲，应考虑把通用作图系统合并到大型专用系统中。作图系统因而应该尽可能简单而高效地运行，这样把作图系统合并到应用系统就不用花很多精力。

现在说明通用 CAD 系统包含的工具，这些工具的作用在于允许用户将图形信息输入计算机并归档。原始数据的输入通过将草图数字化而完成。该系统也允许用户存取、加工、处理并以硬拷贝形式输出这些数据，或者重新归档为永久性存储文件。

使用 CAD 的原因有很多，最有力的推动力就是竞争。为了赢得业务，公司使用 CAD 可以创造出更好的设计，并且在设计速度上比竞争对手更快，在成本上花费更少。通过使用 CAD，生产率得到了很大提高，使用户能够很容易地画多边形、椭圆、多条平行线和多条平行的曲线。在绘制对称部分时，复制、旋转、镜像这些工具使用起来也是很方便的。很多飞机舱口的样式就是用 CAD 程序设计的。用各种不同的颜色填充空白的区域是艺术和表达的需要。CAD 总是提供许多不同类型的字体。能够将不同的图形文件格式和扫描材料（照片）导入 CAD 也是一大优点，特别是可以对图像进行加工、润饰和加入动画效果。

CAD 系统另外一个优点是能够存储在绘图中经常用到的实体。常用零件库可以另外购买或者由绘图员自己创建。在绘图中反复使用的一个典型的项目可以在数秒内检索并确定它的位置，也可定位在任一角度，以满足特定的要求。

使用 CAD 产品，可以通过插入现有的零件图到装配图中，然后按照要求把它们放在合适的位置来绘制装配图。

不同零部件之间的间距能够在图中直接测量。如果需要，可以使用装配图设计出额外的零部件作为参考。

CAD 非常适合文件的快速归档。以前，工程师和绘图员们浪费大约 30% 的时间去寻找图纸和其他文档。用 CAD 产品可以快速而简便地编辑图样，对以前的内容进行修改，更新零件明细表。

当你用纸绘图而客户希望修改图样的时候，你就得全部重画；使用 CAD，可以马上进行修改，并在几分钟之内打印出新图，或者通过电子邮件和互联网立即传送到世界的各个地方。在纸上绘制复杂的几何图形时，经常要进行很多测量并且需要确定参考点；在 CAD 中，这是一件轻而易举的事情，修改也更容易了。许多 CAD 程序包含“宏”或者允许用户定制的附加程序语言。

定制 CAD 系统来使它适合你的特定要求，并用它实现你的天才创意，从而能使你的 CAD 系统区别于你的竞争对手。CAD 能够使企业完成更出色的设计，而用手工的方式几乎是不可能的，同时排除了概念设计阶段的不确定项。

许多 CAD 系统允许快速生成提出的设计模型作为一个线框模型。在 CAD 里创建的实体造型可以输入到一个有限元分析（FEA）程序内，以检验设计方案能否承受预期的负荷。

CAD 总是尽可能地和 CAM（计算机辅助制造）联系在一起。

CAD/CAM 系统能够为计算机化的机床控制器产生计算机指令。例如，车床、磨床、加工中心、旋转冲头、焊接设备和自动化装置等。

CAM 是从数控机床技术中发展来的。早期的数控机床用于伺服驱动装置和电动机的机载电子控制系统，编程要用穿孔纸带，然后及时地变成等效的 ASCII 码文本数据控制流，输入到文本编辑器。

每个设备制造商都开发自己的控制码方案，通常是一组用于表示机械运动的含义隐晦的字符和表示速度、深度等和位置坐标的数值。

数控机床包括一个带有屏幕和键盘的计算机。它们使用“常规”的控制语言。现代的 CAD/CAM 系统能够自动地从三维模型中产生刀具路径，并且能够在屏幕上模拟切削动作。大部分的 CAD/CAM 系统是模块化的，这就意味着用户可以只购买需要的模块并把它们集成为一个统一的系统。

CIM（计算机集成制造）是指基于计算机化信息的制造业各方面的完全集成。

CIM 是 CAD 产生的零件数据在 CAM 环境中的应用。换句话说，就是将计算机化的零件加工的几何形状用于数控编程。这个发展阶段可以称为小规模集成。

CIM 开发的最高级模式是创建一个数据库，其中包含了工厂用于生产零件的柔性制造系统需要的所有信息。在这种模式下，任何需要的人都可以检索和使用。柔性制造是指能够快速、经济地生产任意小批量或大批量的零件，从而减少加工费用、工作量并降低高昂的库存费用。

计算机集成制造中的主要信息流程已经由赫尔伯格清楚地概括出来。CAD 生产产品模型以及产品描述数据，这些数据由 CAPP（计算机辅助工艺规划）转换成工艺流程和 CAM 系统的控制程序。PPC（生产计划和控制系统）系统产生并管理所有在 CAM 中用于控制的操作数据。CAQ（计算机辅助质量保证）校正制造过程中的短期偏差，但从长远看，则影响产品的开发和关于质量保证的方法。

Helberg 的信息流程没有包括系统之间必要的或者期望的更进一步的信息连接。例如，在设计和确定工艺流程时，伴随着计算所进行的 CAD/CAPP 和 PPC 之间的连接，或者从制造到计划的反馈。此外，至少在加工单件零件的情况下，诸如设计和工艺规划这样的过程可以认为是订单交付周期的一部分。因此，它们可以像实际加工和装配过程一样由 PPC 系统规划和控制。在这种情况下，相应的反馈就变得非常必要。

CIM 的理论家们认为，系统集成化和合理化不是一个简单的技术方法问题。为了把分散的不兼容系统集成起来，不能只是简单地投入大量的经费和设备，而必须改变传统的工艺过程。但只要改变工艺过程，就会遇到阻力。公司越大，网络就越独立，改变生产策略和工艺过程就越困难。如图 16.3 所示，成功的系统集成条件是在分布式环境的全局上，沿功能轴和支持轴（硬件和软件）扩展。

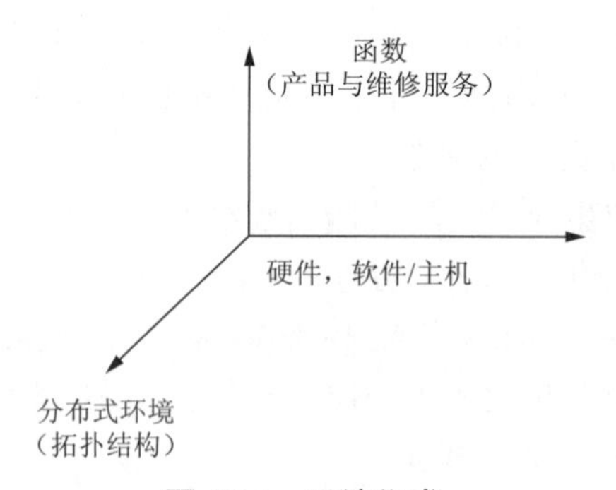

图 16.3 系统集成

因为使用 CIM 带来的切实利益和潜在的利益是长期的，常用的折扣现金收支流和投资回报率的评价方法不能经常有效地评估一个安装了 CIM 的柔性制造系统。相反，必须用战略性优势和潜在的利益来估量对 CIM 投资的期望。

Unit 17

Engineering Drawings

Engineering drawing is a graphical language used by engineers and other technical personnel associated with the engineering profession. The purpose of engineering drawing is to convey graphically the ideas and information necessary for the construction or analysis of machines, structures, or systems. [1] In Figure 17.1, it is shown the operation interface of engineering drawing.

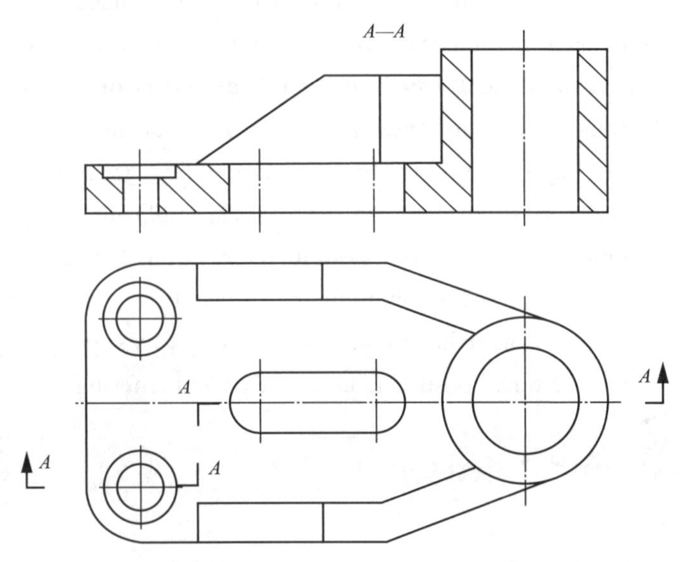

Figure 17.1 Engineering drawing

The basis for many engineering drawing is orthographic representation. Object are depicted by front, top, side, auxiliary, or oblique views, or combinations of these. [2] The complexity of an object determines the number of views shown. At times, pictorial views are also shown.

Engineering drawings often include such features as various types of lines, dimensions, lettered notes, sectional views, and symbols. They may be in the form of carefully planned and checked mechanical drawings, or they may be freehand sketches. Usually a sketch precedes the mechanical drawing.

Many objects have complicated interior detail which cannot be clearly shown by means of front, top, side, or pictorial views. Section views enable the engineer to show the interior detail in such cases. Features of section drawings are cutting-plane symbols which show where imaginary cutting planes are passed to produce the sections, and section-lining, which appears in the section view on all portions that have been in contact with the cutting plane. [3]

If the plane cuts entirely across the object, the section represented is known as a full section.

If it cuts only halfway across a symmetrical object, the section is half section. A broken section is a partial one, which is used when less than a half section is needed. When only a part of the object is to be shown in section, conventional representation such as a revolved, rotated, or broken-out section is used. Thus, certain engineering drawing will be combinations of top and front views, section and rotating views, and partial or pictorial views.

In addition to describing the shape of objects, many drawings must show dimensions, so that workers can build the structure or fabricate parts that will fit together. This is accomplished by placing the required values along dimension lines. In metric dimensioning, the basic unit may be the meter, the centimeter, or the millimeter, depending upon the size of the object or structure.

Types of Working drawings may differ in styles of dimensioning, lettering, positioning of the numbers, and in the type of fraction used. [4] If special precision is required, an upper and a lower allowable limit are shown. Such tolerance or limit dimensioning is necessary for the manufacture of interchangeable mating parts, but unnecessarily close tolerance are very expensive.

A set of working drawings usually includes detail drawings of all parts and an assembly drawing of the complete unit. Assembly drawings vary somewhat in character according to their use, as design assemblies or layouts, working drawing assemblies, general assemblies, installation assemblies, and check assemblies. [5] A typical general assembly may include judicious use of sectioning and identification of each part with a numbered balloon. Accompanying such a drawing is a parts list, in which each part is listed by number and briefly described; the number of pieces required is stated and other pertinent information given. Parts lists are best placed on separate sheets and typewritten to avoid time-consuming and costly hand lettering.

New Words and Phrases

convene *vt.* 召集
vi. 聚集，集合
orthographic *adj.* 正交的
depict *v.* 描述，描绘
auxiliary *adj.* 附加的，辅助的
oblique *adj.* 间接的，斜的
precede *v.* 在……之前；先于
imaginary *adj.* 想象的，虚构的
revolve *v.* 旋转
broken-out section 破断面
fabricate *v.* 制造，装配
metric *adj.* 米制的
interchangeable *adj.* 可互换的
judicious *adj.* 贤明的，判断正确的
pertinent *adj.* 相关的，恰当的

Analyze the Following Sentences

[1]The purpose of engineering drawing is to convey graphically the ideas and information necessary for the construction or analysis of machines, structure, or systems.

工程制图的目的是以图形的方式传达机械、结构或系统的构造或分析所需的想法和信息。（注：graphically 意思为“生动地、轮廓分明地”，用来形容 convey。）

[2]Object are depicted by front, top, side, auxiliary, or oblique views, or combinations of these.

物体由主视图、俯视图、侧视图、辅助视图、斜视图或者这几种视图的组合来描述。（注：are depicted by 意思为“描述了”，此处可理解为“由……描述”。）

[3]Features of section drawings are cutting-plane symbols, which show where imaginary cutting planes are passed to produce the section, and section-lining, which appears in the section view on all portions that have been in contact with the cutting plane.

剖面符号显示想象的剖面在哪里通过而产生剖视图，剖面线出现在剖视图中与剖面接触的所有部分。（注：which 从句结构，注意所修饰的对象，be passed to 意思为“被传递到”。）

[4]Types of working drawings may differ in styles of dimensioning, lettering, positioning of the numbers, and in the type of fraction used.

施工图可以在尺寸标注、文字标注和数字标注位置的风格上，以及所用的分数类型上有所差别。（注：in styles of 意思为“在……风格上”。）

[5]Assembly drawings vary somewhat in character according to their use, as: design assemblies or layouts, working drawing assemblies, general assemblies, installation assemblies, and check assemblies.

根据用途，不同的装配图有所区别，如设计装配或布局、施工图装配、总装配、安装装配和检查装配。（注：vary somewhat 意思为“有些不同”。）

Translation

工程制图是与工程专业相关的，由工程师及其他技术人员使用的一种图形语言。工程制图的目的是以图形的方式传达机械、结构或系统的构造或分析所需的想法和信息。在图 17.1 中，显示了工程图的操作界面。

正视表示法是许多工程制图的基础。物体由主视图、俯视图、侧视图、辅助视图、斜视图或者这几种视图的组合来描述来。物体的复杂程度决定了表达的视图数量,有时也给出示意图。

工程制图通常包括这样的元素，如各种线型、尺寸、文字标注、剖视图和符号。它可以是经过详细计划和检查的机械制图的形式，或者是徒手画的草图的形式。在机械制图前通常是草图。

许多物体有复杂的内部细节，无法用主视图、俯视图或者示意图来清晰地表达。在这种情况下,工程师可以用剖视图表现内部细节。剖面符号显示想象的剖面在哪里通过而产生剖视图,剖面线出现在剖视图中与剖面接触的所有部分。

如果剖面将物体完全切开，被表示的剖面称为全剖视；如果仅切开对称物体的一半，则称为平剖视。局部剖视是部分剖视，当所需剖面少于半个剖面时使用。当只有物体的一部分需要用视图表达时，则使用传统的表达方法，如旋转剖视、旋转视图或切面视图。这样，某些工程图就会由俯视图、主视图、剖面图和旋转视图以及局部视图或示意图组合而成。

除了描绘物体的形状，许多图必须表明尺寸，以便工人可以制作相配合的零部件。这一工作通过沿尺寸线标注所需数值来完成。在米制尺寸标注中，基本单位可以是米、厘米或者毫米，这取决于物体或结构的尺寸。

施工图可以在尺寸标注、文字标注、数字标注位置的风格上，以及所用的分数类型上有所差别。如果需要特殊精度，则要标明允许的上下极限值。这种公差或极限的尺寸标注对于制造可以互换的相匹配的零部件是有必要的，但是没有必要的精密公差代价很高。

一组施工图经常包括所有部件的详图和整个装置的装配图。根据用途，不同装配图有所区别，如设计装配或布局、施工图装配、总装配、安装装配和检查装配。典型的总装配图包括剖切的正确使用和每个零件的识别，零件是用一个标有圆圈的号码来识别的。这种图附有零件明细表，表中的每个零件按号码排列并加以简要叙述；注明所需要的零件号并给出其他相关信息。零件明细表最好放在单独的一张纸上打印出来，以免费时费力的手写。

Unit 18

Wisdom Factory

In recent years, China has constantly introduced targeted measures to promote the development of intelligent manufacturing, and pointed out the new direction of development for the traditional manufacturers—wisdom factory (as shown in Figure 18.1 and Figure 18.2). Wisdom factory will build the communication bridge between a product and the manufacture, for undertaking the implementation of intelligent manufacturing. Therefore, the future of the intelligent manufacturing will be the wisdom factory.

The so-called wisdom factory set a variety of emerging technologies and intelligent systems in one of human chemical plants based on the digital factory. Wisdom factory has improved the control of the production process, reduced human intervention of the production line, timely and accurately collected the operational data, therefore it can enhance the core competitiveness, improve production efficiency and reasonably arrange production and so on.

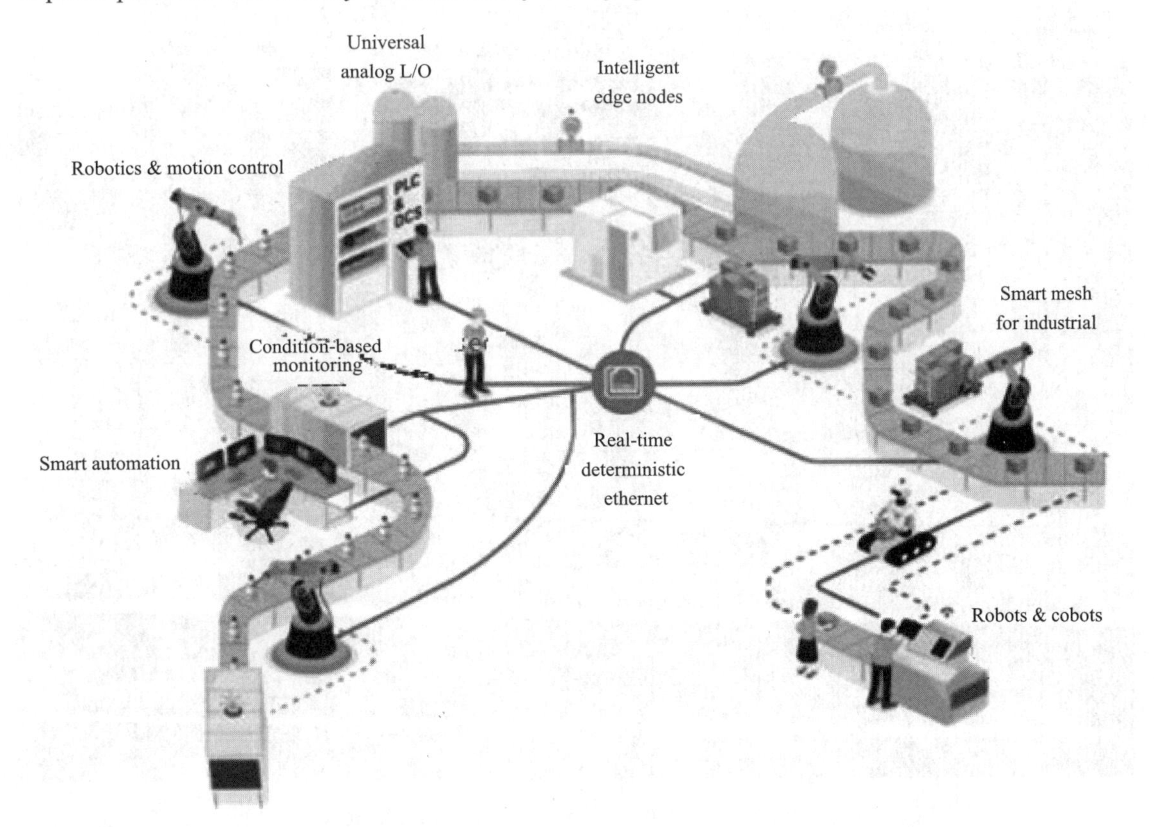

Figure 18.1 Wisdom factory architecture

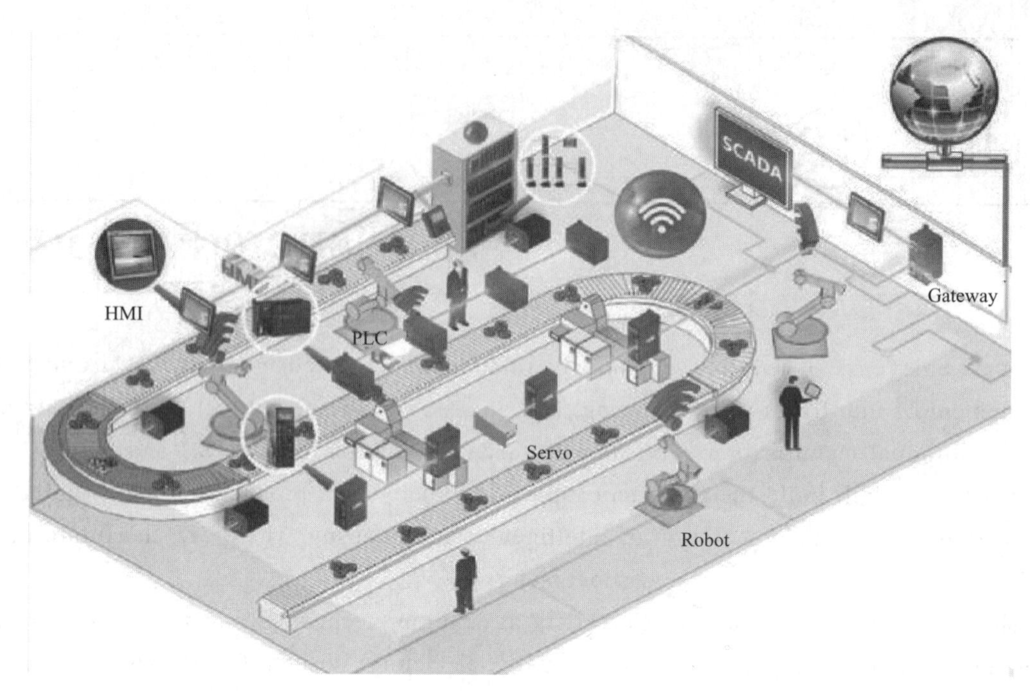

Figure 18.1 Wisdom factory architecture (continued)

Figure 18.2 An example of an intelligent factory

From the definition we can see, the realization of intelligent factories can't be separated from the support and application of emerging technologies. For example, without the extensive use of advanced sensors, the wisdom factory is difficult to be called smart. The development of advanced sensors relies on the progress of microprocessors and artificial intelligence technology.

The application of software engineering assistance system is also one the basic elements of intelligent factory. Software engineering assistance system is a based on knowledge, highly integrated intelligent software system, with digital expression, information access, processing knowledge and other capabilities. The wisdom factory will gradually replace the traditional single mode of work.

Of course, to realize the wisdom factory, we need to focus on breaking through machine tools, processes, production control and other key technologies.

Take machine tools as an example, CNC machine tools are the most basic part of the wisdom factory. Therefore, the construction of the wisdom factory can't be separated from the advance of the intelligent machine tools. Intelligent machine tools can independently collect data and determine the operation status, automatically detect, induce, simulate the intelligent decision-making of the goal, so that the machine operation is in the best condition. [1]

In addition, the development of intelligent logistics is also a key factor of the fact that intelligent factories rapidly spread. For intelligent factories, intelligent logistics is a "recycling system" which needs continuously to transport production-related resources. [2] Fortunately, China vigorously developed intelligent logistics in recent years; information and intelligent construction have made great progress.

In general, the wisdom factory is the ultimate goal of the modern plant being information technology development, and also an important step to achieve intelligent manufacturing.

New Words and Phrases

targeted initiative 有针对性的措施

wisdom factory 智慧工厂

human chemical plant 人性化工厂

production line 生产线

core competitiveness 核心竞争力

artificial intelligence 人工智能

software engineering 软件工程

machine tool 机床

production control 生产控制

recycling system 回收系统

Analyze the Following Sentences

[1] Intelligent machine tools can independently collect data and determine the operation status, automatically detect, induce, simulate the intelligent decision-making of the goal, so that the machine operation is in the best condition.

智能机床可以独立收集数据并确定运行状态，自动检测、诱导、模拟目标的智能决策，使机器运行处于最佳状态。（注：independently 意思为“独立地”。）

[2] For intelligent factories, intelligent logistics is a “recycling system” which needs continuously to transport production-related resources.

对于智慧工厂而言，智能物流是一个“回收系统”，需要持续运输与生产相关的资源。（注：recycling 是 recycle 的现在分词，意思为“回收利用、再利用”。）

Translation

近年来，我国不断出台针对性举措推动智能制造的发展，为传统制造型厂商指出了新的发展方向——智慧工厂（见图 18.1、图 18.2）。智慧工厂的提出，将搭建起产品和制造之间的沟通桥梁，为智能制造起到承接落地作用。就因此，智能制造业的未来将是智慧工厂。

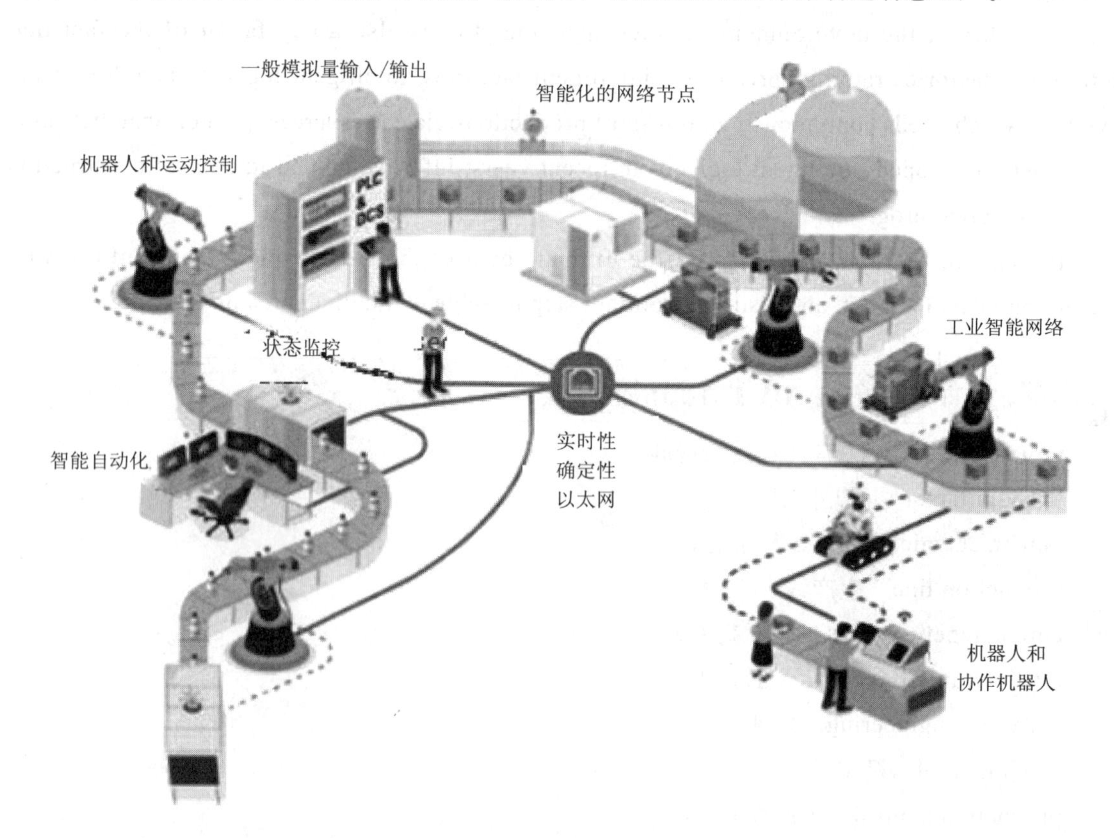

图 18.1　智能工厂结构

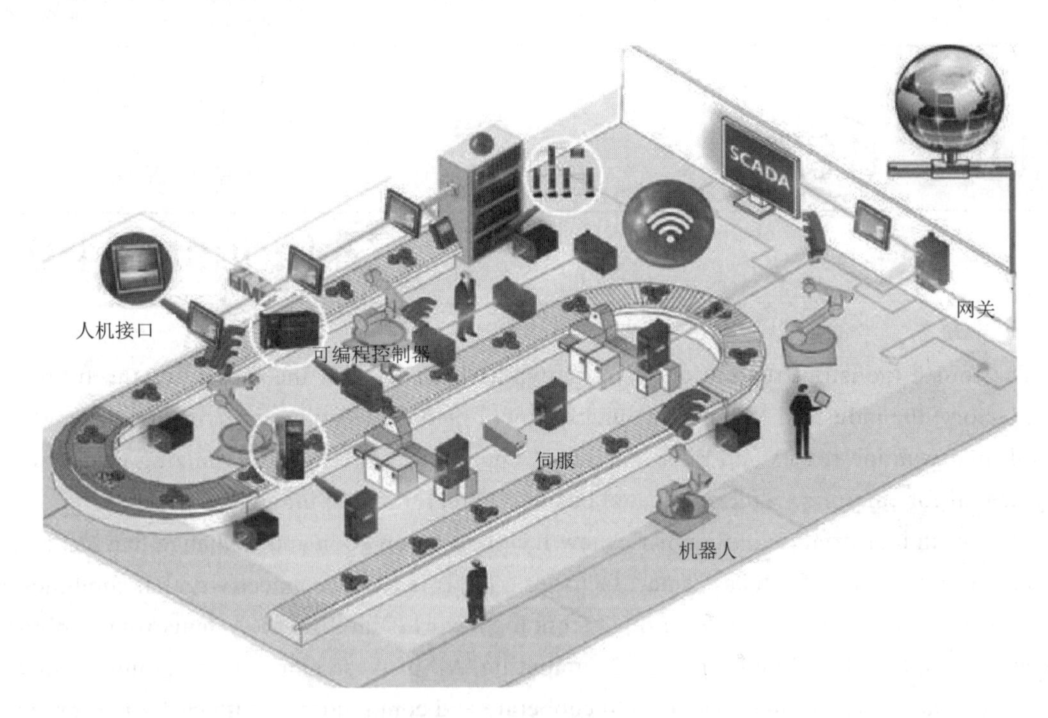

图 18.1　智能工厂结构（续）

所谓智慧工厂，就是在数字化工厂的基础上，集各种新兴技术和智能系统于一体而构建的人性化工厂。智慧工厂提高了生产过程的可控性，减少了生产线的人工干预，及时准确地采集作业数据，因此能增强核心竞争力、提高生产效率及合理安排生产等。

从定义可以看出，智慧工厂的实现离不开新兴技术的支持与应用。例如，没有先进传感器的广泛应用，智慧工厂难以称得上智能。而先进传感器的发展依赖于微处理器和人工智能技术的进步。

软件工程辅助系统的应用也是智慧工厂的基本构成要素之一。软件工程辅助系统是一个基于知识、高度集成的智能软件系统，具有数字表达、信息访问、知识处理等能力。智慧工厂将逐步取代传统的单一工作模式。

当然，为真正实现智慧工厂，还需要重点突破机床、工艺、生产控制等关键技术。

以机床为例，计算机数控机床是智慧工厂最基本的部分。因此，智慧工厂的建设离不开智能机床的进步。智能机床可以独立收集数据并确定运行状态，自动检测、诱导、模拟目标的智能决策，使机器运行处于最佳状态。

此外，智能物流的发展也是智慧工厂快速发展的关键因素。对于智慧工厂而言，智能物流是一个“回收系统”，需要持续运输与生产相关的资源。幸运的是，我国近年来大力发展智能物流，信息化、智能化建设取得了长足进展。

总的来说，智慧工厂是现代工厂信息化发展的最终目标，也是实现智能制造的重要一步。

Unit 19

Industry 4.0

The fourth industrial revolution, also known as Industry 4.0, the factory of the future, the smart factory, the industrial Internet, is quickly emerging and affecting our lives in new ways. It is helpful to streamline robots and their automated operations, while also optimizing costs, turning the potential for enterprise-wide automated business transformation into a reality.

The fourth industrial revolution is the new fusion of automation and exchange that has created the perfect environment for these "smart factories". [1] It is a new connectivity that combines and networks all of the advancements in our technology world: intelligent systems (cyber-physical systems), the Internet of things, and cloud computing. All of these items are programmed to work together to enable humans and machines to cooperate and communicate, with each other or among themselves, in real time and even remotely. All of these items are programmed to work together to enable humans and machines to cooperate and communicate, with each other or among themselves, in real time and even remotely. All of the devices, systems, and people in the production chain are linked and able to deliver data in the right form whenever and wherever necessary.

Industry 4.0 is continuing to prove that humans are unstoppable in their innovative solutions towards a more effective and intelligent society. Since the beginning, humans have been on a constant search to continue innovating and finding new solutions that make creating products more affordable, smarter, and also make our lives a little easier. [2] The revolutions of the past (water and steam power, electric power, and digital power) have truly transformed the society, governing structures, and human identities in which they are found. All of these individual revolutions have aided manufacturing businesses to continue their unvarying searches for better and more economically feasible solutions.

The diagram of four industrial revolution is shown in Figure 19.1.

(1) The first industrial revolution (or steam technology revolution) took place from the 18th to 19th centuries in Europe and America.

(2) The second industrial revolution (or power technology revolution) took place between 1870 and 1914, just before World War I.

(3) The third revolution of science and technology (or information technology revolution)was marked by the invention and application of atomic energy, electronic computer, space technology and biological engineering.

(4) The fourth industrial revolution was marked by Internet industrialization, industrial intelligence and industrial integration, mainly including artificial intelligence, clean energy, unmanned control technology, quantum information technology, virtual reality, and biotechnology.

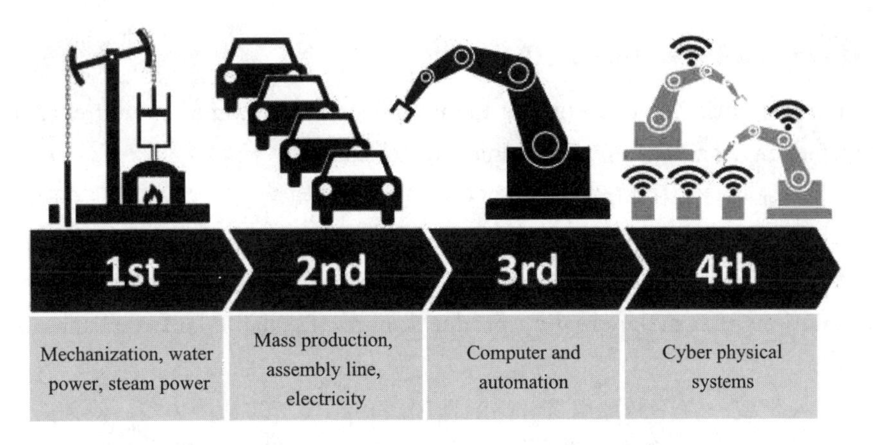

Figure 19.1 The diagram of four industrial revolution

The new industrial revolution will bring new momentum to global economic growth and promote changes in human production and lifestyles.

So the industrial revolution we see today is really of no surprise. Reliant on all revolutions that existed before it, the fourth industrial revolution is offering entirely new capabilities for people and machines and ways in which technology becomes embedded within societies. The level of manufacturing agility can make it possible to connect customers' needs with the ability of company to deliver a product, potentially on demand. Manufacturers will be able to listen and then act to better adapt to the consumers' demands as they can access real-time intelligence and analytics, on-demand.

If we have the courage to take collective responsibility for the changes underway, and the ability to work together to raise awareness and shape new narratives, we can embark on restructuring our economic, social systems to take full advantages of emerging technologies.

It is clear that the fourth industrial revolution is here to stay and if you are in the manufacturing business, sitting on the sidelines is no longer a good idea. Robotic technology is well on its way to revolutionize the way to the success of a business; the benefits are increasingly clear and attainable to those of all sizes.

New Words and Phrases

industrial revolution 工业革命
factory of the future 未来工厂
smart factory 智慧工厂
industrial Internet 工业互联网
intelligent system 智能系统
Internet of things 物联网
cloud computing 云计算
real time 即时的

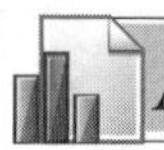

Analyze the Following Sentences

[1] The fourth industrial revolution is the new fusion of automation and exchange that has created the perfect environment for these “smart factories”.

第四次工业革命是自动化和数据交换的新融合，为“智慧工厂”创造了完美的环境。(注：revolution 意思为“革命”；fusion 意思为“融合”。)

[2] Since the beginning, humans have been on a constant search to continue innovating and finding new solutions that make creating products more affordable, smarter, and also make our lives a little easier.

从一开始，人类就一直不断地研究，不断地创新和寻找新的解决方案，使创造产品更实惠、更智能，也使我们的生活更容易一些。(注：innovating 意思为“创新、改革”。)

Translation

第四次工业革命，又称工业 4.0、未来工厂、智慧工厂、工业互联网，正在迅速涌现，并以新的方式影响我们的生活。它有助于简化机器人及其自动化操作，同时还可以优化成本，将企业广泛的自动化业务转型的潜力变成现实。

第四次工业革命是自动化和数据交换的新融合，为“智慧工厂”创造了完美的环境。它是一种将技术领域中所有进步都结合和联网的新的连接，如智能系统（网络物理系统）、物联网和云计算。所有的物品被编程为一起工作，使人和机器能够实时甚至远程地进行合作和通信。生产链中的所有设备、系统和人员都被连接到一起，并能够随时随地地以正确的形式传递数据。

工业 4.0 正在不断地证明，人类正在不可阻挡地向着更加有效和智慧的社会方向创新。从一开始，人类就一直不断地研究，不断地创新和寻找新的解决方案，使创造产品更实惠、更智能，也使我们的生活更容易一些。过去的革命（水和蒸汽动力、电力和数字动力）真正改变了社会、管理结构和人的特性。所有这些革命都帮助制造业不断寻找更好、更经济可行的解决方案。

四次工业革命一览图如图 19.1 所示。

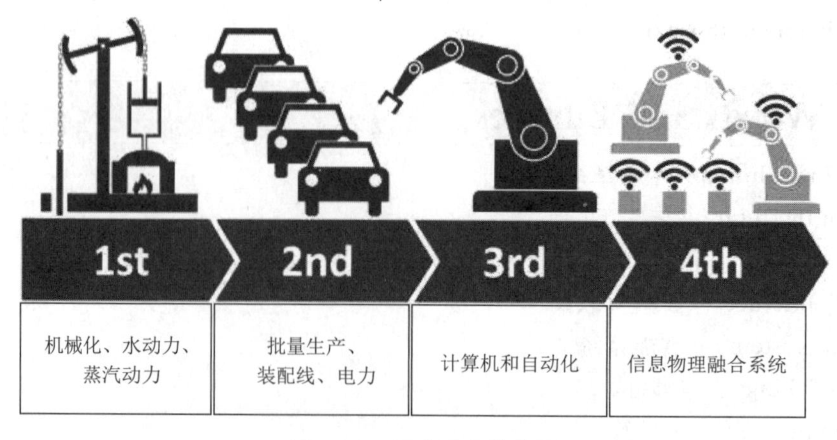

图 19.1　四次工业革命图

（1）第一次工业革命（或蒸汽技术革命）发生于18世纪至19世纪的欧洲和美国。

（2）第二次工业革命（或动力技术革命）发生于1870年至1914年，就在第一次世界大战之前。

（3）第三次科技革命（或信息技术革命）的标志是原子能、电子计算机、空间技术和生物工程的发明和应用。

（4）第四次工业革命的标志是互联网工业化、工业智能化和产业一体化，主要包括人工智能、清洁能源、无人控制技术、量子信息技术、虚拟现实和生物技术。

新的工业革命将给全球经济增长带来新的动力，促进人类生产和生活方式的改变。

因此，我们现在看到的工业革命真的是不足为奇。依靠之前发生的所有革命，第四次工业革命正在为人员和机器以及将技术嵌入社会的方式提供全新的功能，制造敏捷性的水平可以使客户的需求与公司根据潜在需求提供产品的能力联系起来。制造商将能够进行倾听，然后采取行动以更好地适应消费者的需求，因为他们可以按需访问实时信息并分析。

如果我们有勇气为正在进行的变革承担主要责任，共同努力提高认识和形成新的规则，我们就可以重新调整经济、社会制度，充分利用新兴技术。

很明显，第四次工业革命已经到来，如果你在制造业，袖手旁观不再是一个好主意。机器人技术正在彻底改变企业成功的方式，其利益越来越明显，可以达到各种规模。

Reference

[1] 姚薇，李瑞年 . 电气自动化技术专业英语 [M]. 北京：中国铁道出版社，2013.
[2] 陈春丽 . 电气工程专业英语 [M]. 北京：机械工业出版社，2008.
[3] 张锦，王婷 . 电气自动化技术英语教程 [M]. 北京：中国铁道出版社，2011.
[4] 王磊，涂杰 . 机电工程专业英语 [M]. 北京：冶金工业出版社，2009.
[5] 张明文 . 工业机器人专业英语 [M]. 武汉：华中科技大学出版社，2018.
[6] 肖伟平，夏龙军 . 电梯专业英语 [M]. 北京：化学工业出版社，2012.